I0019579

WAP and WML: Designing Usable Mobile Sites

Ryan Younger

Table of Contents

Chapter 1

1.1 – Introduction – WAP, why all the fuss?

WAP or Wireless Application Protocol is without a doubt one of today's buzzwords, WAP is a technology that allows its users to access WAP pages or 'cards' on the Internet. There was a great deal of hype in the media when WAP technology was first introduced in the UK. Since that time however only a small minority of companies have taken the step of actually setting up a WAP site, yet according to IDC Research (2001) the 7 million WAP phones in Western Europe at the start of 2001 is set to rise to around 50 million by 2004.

IDC Research (2001) also indicates that 80% of the UK's population will own and regularly use an Internet-enabled phone by 2005. Amongst the small minority of British companies currently operating WAP sites are Barclays Bank who on Wednesday May 23, 2001 launched (in co-operation with Vodafone) what they claim is the UK's first WAP service to combine banking, credit card and share dealing services. Barclays customers can use the service to check their balance, pay bills and get directions to the nearest cash machine (Guardian Newspaper, 2001).

Barclays is not the first bank to offer a WAP service, however it is the first to implement such powerful features within the site. With Barclays Bank taking the lead in offering powerful features to customers with access to a WAP phone it is likely that other companies will follow suit as was witnessed with the growth of the World Wide Web in the UK. Some feel the technology is doomed to failure, whilst others feel it will drive M-Commerce (Mobile Commerce) and provide a means for users to access a large number of products and services via their mobile phones on the move.

1.2 – What will you get from this book?

This book will cover WAP and WML basics, WAP development packages, the prototype evaluation process, user testing, the viability of WAP within businesses, WAP usability best practices and will also provide a selection of real world WAP case studies for you to read through at your leisure. I hope you find this book a useful and enjoyable read.

Chapter 2

2.1 – What is WAP?

WAP stands for Wireless Application Protocol and is a secure specification that allows users to access information instantly via handheld wireless devices such as mobile phones, pagers, two-way radios, smartphones and communicators. WAP Forum, the creators of the WAP protocol define WAP as 'The de facto standard for providing communications and advanced telephony services on digital mobile phones, pagers, personal digital assistants and other wireless terminals'. A common misunderstanding regarding WAP is that you can 'surf the web' just as on a desktop PC. The truth is that most WAP phones cannot simply render a web page and filter the contents into a text-based display. In order for HTML pages to be viewed by a WAP enabled phone they need to be rewritten in WAP's proprietary language named WML or Wireless Markup Language.

2.1.1 – WAP Gateway Model

As with the Internet, WAP has clients and servers too. The mobile device, usually a WAP phone acts as the client, whilst standard Internet servers hold static WAP pages and applications. However the wireless device has to connect to the server without any wires so that the device can make a request to the server for a page. In order to effectively connect the device and the server, WAP specifications state that there must be a WAP gateway. The gateway converts the WAP page request (WML) into a web page request (HTML) and the returning Web response (HTML) into a WAP response (WML) again. Figure 1 demonstrates how this process works.

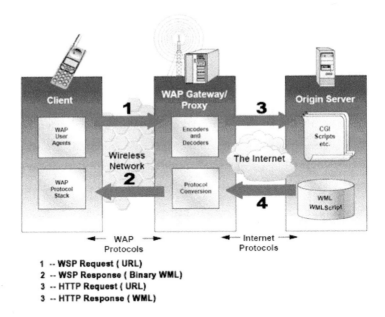

1 -- WSP Request (URL)
2 -- WSP Response (Binary WML)
3 -- HTTP Request (URL)
3 -- HTTP Response (WML)

Figure 1. WAP gateway model

2.1.2 - WAP Protocol Stack

WAP utilises a communications stack protocol similar to that used by the web. The protocol stack covers both the application and transport layers. As WAP is to work on communications devices that have low-bandwidth, high latency and lossy characteristics the stack is designed to be scalable, minimise the bandwidth and maximise the number of wireless network types that can deliver WAP content (Figure 2.).

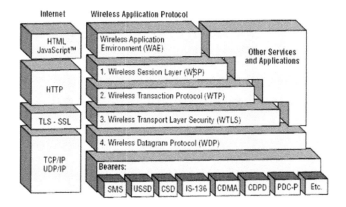

Figure 2. WAP protocol stack

At the application layer, WAP uses WML and WMLscript to produce WAP content. For the transport layer WAP uses:

- WDP (Wireless Datagram Protocol)
- WTLS (Wireless Transport Layer Security)
- WTP (Wireless Transaction Protocol)
- WSP (Wireless Session Protocol)

2.1.3 – Wireless Markup Language (WML)

Traditional web pages are written in HTML or Hyper Text Markup Language whilst pages designed for a WAP phone are written in WML. WML has been specifically designed for small handheld devices with low memory and processing constraints that utilise the low-bandwidth of a wireless-handheld network. Thus WML pages are, the majority of the time, text based and without any colour, sound or animations.

The main purpose of WML is to describe the content and format of a page; this is done effectively due to the fact that it is a subset of the content-tagging language XML. However for more advanced operations such as searching an on-line database a wireless version of JavaScript known as WMLscript would be utilised.

2.2 – Wireless Markup Language and HyperText Markup Language

The majority of websites on the Internet today are written in HTML, the reason HTML is so popular is that it is very versatile. With it a coder can produce the amazing range of layouts we see on the web today. HTML has evolved into the language we see today over time, the fact that it is so widespread means that most browsers are forgiving of coding errors and will attempt to display more or less any HTML code. WML on the other hand is less forgiving due to the fact that it is designed for use across the maximum number of portable devices with the minimum number of compatibility issues. This results in WML being a far stricter language than HTML. A browser immediately reads HTML code whereas WML code needs to be compiled by a 'WAP gateway' before the WAP devices mini browser can display it. If the WML code has errors then it will not be compiled by the gateway and in turn will not be able to be viewed on the WAP device attempting to access the page.

2.2.1 – Main differences between WML and HTML

All WML tags are lower case while HTML tags can be a mixture of cases, meaning that upper case, lower case or even a mixture of cases are acceptable. WML tags must always be lower case without exception.

A WML page is known as a 'deck' as it can have several sub-pages within it; these subpages are referred to as 'cards'. This allows navigation to be far quicker as a deck is downloaded to the phones internal memory whilst a particular card is being viewed, meaning that when the user wishes to view the next card in the deck it will already be in the phones memory.

WML supports multiple <card> tags. The HTML equivalent of a WML <card> tag is the HTML <BODY> tag. However a web page uses only a single <BODY> tag within each page due to the fact that the bandwidth available to a PC accessing the Internet is far higher than that of a WAP device meaning that HTML pages can be downloaded 'on the fly' as and when required.

WML pages have no frames, fonts or colours whilst HTML is media rich. Pages written in WML are not media rich due to the fact that they are to be viewed on small mobile devices with limited capabilities.

2.2.2 – Common WML tags and their purpose

A listing of some common WML tags follows:

Deck & Card Elements

WML Elements	Purpose
<!-->	Defines a WML comment
<wml>	Defines a WML deck (WML root)
<head>	Defines head information
<meta>	Defines meta information
<card>	Defines a card in a deck
<access>	Defines information about the access control of a deck
<template>	Defines a code template for all the cards in a deck

Text Elements

WML Elements	Purpose
 	Defines a line break
<p>	Defines a paragraph
<table>	Defines a table
<td>	Defines a table cell (table data)
<tr>	Defines a table row

<pre>	Defines preformatted text

Text Formatting Tags

WML Elements	Purpose
<b>	Defines bold text
<big>	Defines big text
<em>	Defines emphasized text
<i>	Defines italic text
<small>	Defines small text
<strong>	Defines strong text
<u>	Defines underlined text

Image Elements

WML Elements	Purpose
<img>	Defines an image

Anchor Elements

WML Elements	Purpose
<a>	Defines an anchor
<anchor>	Defines an anchor

Event Elements

WML Elements	Purpose
<do>	Defines a do event handler
<onevent>	Defines an onevent event handler
<postfield>	Defines a postfield event handler
<ontimer>	Defines an ontimer event handler
<onenterforward>	Defines an onenterforward handler
<onenterbackward>	Defines an onenterbackward handler
<onpick>	Defines an onpick event handler

Task Elements

WML Elements	Purpose
<go>	Represents the action of switching to a new card
<noop>	Says that nothing should be done
<prev>	Represents the action of going back to the previous card
<refresh>	Refreshes some specified card variables.

Input Elements

WML Elements	Purpose
<input>	Defines an input field
<select>	Defines a select group
<option>	Defines an option in a selectable list
<fieldset>	Defines a set of input fields
<optgroup>	Defines an option group in a selectable list

Variable Elements

WML Elements	Purpose
<setvar>	Defines and sets a variable
<timer>	Defines a timer

2.2.3 – Hello World WML example

Let's take a look at an example of a typical WML document:

```
<?xml version="1.0"?>
<!DOCTYPE wml PUBLIC "-//WAPFORUM//DTD WML 1.3//EN"
"http://www.wapforum.org/DTD/wml13.dtd">

<wml>
```

```
<card id="card1" title="WML Example">
 <p>Hello World</p>
</card>

<card id="card2" title="WML Example">
 <p>Hello again</p>
</card>
</wml
```

This example will output the following text on the screen of the WAP device:

Hello World

2.2.4 WML Metadata

Metadata can be specified within your WML file using the <meta/> tag. The metadata must be placed between the <head></head> tags as shown below:

```
<?xml version="1.0"?>
<!DOCTYPE wml PUBLIC "-//WAPFORUM//DTD WML 1.3//EN"
"http://www.wapforum.org/DTD/wml13.dtd">

<wml>
 <head>
  <meta name="author" content="Ryan Younger"/>
 </head>

<card id="card1" title="WML Tutorial">
 <p>Hello World</p>
```

```
</card>
</wml>
```

The WAP browser will ignore any metadata if it does not understand its meaning. Metadata of any sort can be specified in your WML file without affecting the look of the card, such as the example above where we include the author's name.

2.2.5 WML Comments

WML comments are placed inside the <!-- --> tags in WML in the same fashion as HTML, comments are ignored by the WAP browser and not rendered on the card. The following example shows an example of WML comments:

```
<!—WML comment -->
<!-- Multi-line
    WML comment -->
```

2.2.6 Setting Font Size and Style

A number of tags are provided within WML to set the font size and style of text rendered on the device. It should be noted that older devices do not support some of these tags, however these will simply be ignored by such devices and will not generate any error messages. Some common font size and style tags include:

```
<?xml version="1.0"?>
<!DOCTYPE wml PUBLIC "-//WAPFORUM//DTD WML 1.3//EN"
"http://www.wapforum.org/DTD/wml13.dtd">

<wml>
```

```
<card id="card1" title="Font Style">
  <p>
        <small>Small</small><br/>
        <big>Big</big><br/>
        <em>Emphasis</em><br/>
        <strong>Strong</strong>
        <b>Bold</b><br/>
        <i>Italic</i><br/>
        <u>Underline</u><br/>
  </p>
 </card>
</wml>
```

The above WML code would generate the following output on the WAP device:

Small

Big

Emphasis

Strong

Bold

Italic

Underline

2.2.7 WML tables

The <table>, <tr> and <td> WML tags are used to create tables. Tables (<table>) have multiple rows (<tr>) and each row has multiple cells (<td>). The following example illustrates the code for a WML table:

```
<?xml version="1.0"?>
<!DOCTYPE wml PUBLIC "-//WAPFORUM//DTD WML 1.3//EN"
"http://www.wapforum.org/DTD/wml13.dtd">

<wml>
```

```
  <card id="page1" title="WML Table Example">
   <p>
    <table columns="3">
     <tr>
      <td>Cell 1</td>
      <td>Cell 2</td>
      <td>Cell 3</td>
     </tr>

     <tr>
      <td>Cell 4</td>
      <td>Cell 5</td>
      <td>Cell 6</td>
     </tr>
    </table>
   </p>
  </card>
</wml>
```

This WML code results in the following output on the WAP device:

Cell 1	Cell 2	Cell 3
Cell 4	Cell 5	Cell 6

2.2.8 WML input fields

The <input/> tag is used to create input fields as the following example illustrates:

```
<?xml version="1.0"?>
<!DOCTYPE wml PUBLIC "-//WAPFORUM//DTD WML 1.3//EN"
"http://www.wapforum.org/DTD/wml13.dtd">
```

```
<wml>
 <card id="card1" title="Input Fields">
  <p>
   Input Field Example.<br/>
   What's your age?
   <input name="exampleinputfield" maxlength="16"/>
  </p>
 </card>
</wml>
```

Our WML code will generate the following output on the WAP device:

Input Field Example.

What's your age?

```
┌─────────────────────────────────────┐
│                                     │
└─────────────────────────────────────┘
```

2.2.9 WML anchor element

Anchor links enable navigation between different WML cards, selecting an anchor link will take you to another WML card in the current, or another deck. The <anchor></anchor> tags are used to create anchor links in conjunction with tags such as the <go/> and <prev/> tags which tell WAP browsers what action to take when the link is selected. The following example illustrates the use of the anchor element:

```
<?xml version="1.0"?>
<!DOCTYPE wml PUBLIC "-//WAPFORUM//DTD WML 1.2//EN"
"http://www.wapforum.org/DTD/wml12.dtd">
<wml>
<card title="Anchor Link">
```

```
<p>
  <anchor>
   <go href="nextpage.wml"/>
  </anchor>
</p>
<p>
<anchor>
   <prev/>
  </anchor>
</p>
</card>
</wml>
```

2.3 - WMLScript

Just as HTML web pages often use JavaScript for more advanced operations, WML pages use their own version of JavaScript known as WMLScript. 'WAP Forum' offers a good working definition of WMLScript:

'WMLScript is the WAP equivalent of JavaScript. More accurately, WML Script is based on JavaScript, uses similar syntax and constructs, and provides semantically equivalent functions.' Wap Forum (2001).

WML Script is fast becoming the standard development platform across all mobile WAP devices. WML Script allows mobile devices to use standard functions across the board as opposed to each device having its own functions or applications. This means that users of different mobile devices can use the same interface and set of applications or functions, independent of brand. WML Script not only allows each mobile device to use standard functions, but also empowers WAP Site developers to implement advanced features such as search facilities and real time site updates. However many professionals within the IT industry have expressed concern over the security risks associated with WML Script.

2.3.1 – Security risks of WML script

According to WAP Forum the developers of WML Script have 'ignored the lessons learned from past security problems with JavaScript and other mobile code technologies. The security risks associated with WML Script are based on a fundamental lack of a model for secure computation'. (WAP Forum, 2001)

A journal article in Communications Of The ACM criticised the fact that the WML script virtual machine does not appear to have a mechanism for preventing access to persistent storage on a device from untrusted scripts. Apparently this results in personal identifying information kept on a device being susceptible to unauthorised disclosure from malicious WML scripts that download and read the information and then ship it off to other sites. (Communications Of The ACM (2001))

This is of great concern as WAP is being hailed as 'the next big thing' in the world of E-Commerce or 'M-Commerce' as it is known in the wireless world. Other attacks include online application duplicity and e-mail virus attacks. The most significant risk to M-Commerce systems will be from malicious code that is beginning to penetrate wireless networks. According to the ACM encrypted communication protocols are necessary to provide confidentiality, integrity, and authentication for M-Commerce applications.

2.4 – Building a WAP site

There are a number of different ways of developing a WAP site depending on your skill level.

2.4.1 – On-line building tools

For absolute beginners there are a number of websites such as Wapdrive (http://www.wapdrive.com) that have on-line Java Based 'builder' scripts which allow you to build your site with no WML knowledge. Wapdrive define their 'Basic Builder' as:

'An easy template system for creating your very own WAP site - visible on any WAP phone worldwide'

Wapdrive.com (2001)

2.4.2 – WAP page editor

For those at an intermediate level there are a great number of WAP Page Editors that are basically the WAP equivalent of GUI based Web Page Editors on PC's such as Microsoft FrontPage or Macromedia Dreamweaver.

2.4.3 – Coding from scratch

For the true veteran, there is the option to code an entire WAP site from scratch using a standard ASCII text editor such as Microsoft Notepad. According to 'WAP.com' there are also a number of editors that allow you to preview your WAP pages whilst coding them such as 'WaPPage' and 'WAPtor'. They also state that you can use a popular Web Page Editor - Macromedia's Dreamweaver if you download a WAP extension. This would be a viable option for those accustomed to using Dreamweaver for web page design and would mean that they would not have to learn a whole new interface.

2.5 – Viewing a WAP site

There are two main ways of viewing a WAP site. The easiest way is of course to use a wireless device such as a WAP phone and to enter the URL (Universal Resource Locator) of the site you wish to access. Another way to view a WAP site is to use an emulator that runs on a desktop PC and allows you to view sites written in WML.

2.5.1 – Using a WAP phone to view a WAP site

The most common way of accessing a WAP site is by using a WAP phone, however a number of other handheld wireless devices can also access WAP sites such as Palm Pilots and a number of Psion Organisers, when connected to a mobile phone. A user accesses a WAP site in much the same way as they would a website by inputting the destination URL of the site required. WAP phones use their own integrated browser software known as a 'microbrowser' to display decks of WML cards.

2.5.2 – Using a desktop PC to view a WAP site

It is possible to view WAP pages on an ordinary Desktop PC with an Internet connection. There are two ways to do this you can either download a WAP emulator and install it on your PC such as WinWap, or you can use an on-line WAP emulator such as the 'Wapalizer' at Gelon.net. According to Gelon.net:

'The Gelon WAPalizer has become recognised as the industry's leading WAP emulator. Based on a substantial number of requests, we have decided to make the WAPalizer available for use on other websites. The WAPalizer is unique in that it actually looks like a real WAP phone (Figure 3). This ensures that the user sees a WAP site, as it would actually appear on the WAP phone. This makes it easy to illustrate the benefits of WAP or showcase a company's WAP site to end-users'.

Gelon.net (2001).

 **Figure 3. Gelon.net Wapalizer**

An emulator allows a WAP site to be viewed on many different types of phones with relative accuracy, which is of great use when developing a WAP site.

2.6 - M-Commerce

2.6.1 – What is M-Commerce

M-commerce is the term used to define the buying and selling of goods and services through wireless devices such as WAP phones. Also known as next-generation E-commerce, M-Commerce allows wireless users to have access to:

- Financial services, this includes mobile banking (customers can use their handset to access their accounts and pay their bills) as well as brokerage services, in which stock quotes can be displayed and traded from a WAP phone.

- Telecommunications, in which bill payment, service charges and account reviews can be conducted from a WAP phone.

- Information services that include the delivery of financial news, sports figures and traffic updates to a WAP phone.

2.6.2 – Problems with M-Commerce

There is currently one major problem with M-commerce in the UK. This is the current standard mobile network, known as 'GSM'. The GSM network is part of the 2nd generation of mobile phones (digital phones); the first generation of mobile phones were analogue. The problem with the GSM network is that it is not optimally suited for mobile-data transfer as the connections are unstable and the data transfer rate is limited to 9.6 kbps. This is extremely low if you wanted to, say, process a transaction on your WAP phone via a WAP site. Due to these technical problems M-commerce has not really taken off yet and as a result actual revenues in mobile commerce are fairly low. However the 3rd generation of mobile phones will be released soon utilising superior networks which should help to alleviate these issues.

2.6.3 - UMTS

UMTS or Universal Mobile Telecommunications System does not use copper wires, therefore it is capable of a bandwidth of up to 2Mbps (35 times faster than a standard 56k dial-up modem in a desktop PC). This kind of bandwidth will allow the use of videoconferencing and video clips., however when UMTS is first released it will only be running at 384 kbps. UMTS should be launched commercially sometime this year and licenses have already been awarded in several European countries. UMTS experimental systems are now in field trials with leading vendors worldwide.

2.6.4 - GPRS

GPRS or General Packet Radio Service is a packet switched transmission technology optimised for multimedia services to wireless devices. One advantage of this technology is that users are always connected meaning they are always on-line but can only be charged for the amount of data that is transported. Users will benefit from fast and easy access to 170 kbps data transfer rates. The main benefits of GPRS are:

- Cost reduction due to volume dependent charging.
- Higher user data rates.
- Faster Internet access.
- New applications enabling real plug and play.
- Prepaid services also possible for data.

2.7 – Rival technology: i-mode

i-mode is a rival technology to WAP currently operational in Japan, i-mode allows users to browse traditional (HTML) web pages (with varying degrees of success) as well as pages written in i-mode's proprietary language. The i-mode mobile Internet access system was created by Japanese company 'DoCoMo', the technology is 'packet-switched' (using Japans Personal Digital Cellular Packet (PDC-P) network) which means that i-mode is in principle 'always on', provided the user is in an area where i-mode reception is available. There are currently over 16 million i-mode subscribers, increasing by about 50,000 per day. According to DoCoMo's data, the large majority of subscribers use e-mail and browse web pages every day. DoCoMo are planning to bring i-mode to Europe as soon as the required networks are available (GPRS) making i-mode a potential future threat to WAP in the UK and the rest of Europe.

2.8 – The WAP forum

The WAP Forum is an industry association that created the WAP specification and continues to release all future specifications. The WAP Forum comprises over 500 members that have developed the de-facto world standard for wireless information and telephony services on digital mobile phones and wireless terminals. WAP Forum members represent over 90% of the global handset market, carriers with more than 100 million subscribers, leading infrastructure providers, software developers and other organisations providing solutions to the wireless industry.

Notable members of the WAP Forum include leading manufacturers Nokia and Ericsson.

2.9 – Benefits of WAP

WAP technology offers a number of benefits including access to important information such as news reports, email, on-line shopping and even on-line multiplayer games. All six of the phone networks in Britain have now launched WAP phones and dedicated WAP services, with each possessing their own portal and pools of information. The WAP version of Lastminute.com allows you to search for bargain holidays and cheap flights and even to book a table at a London restaurant. WML versions of Yahoo! and Excite provide customisable on-line organisers and diaries, while the Guardian newspaper complements its website with a WAP-specific selection of news, commentary and reviews. Advanced mobile technology and mobile Internet connectivity is the future of computing and WAP technology is a big leap toward this.

CHAPTER 3

3.1 – Planning a WAP site

'The principles that have informed quality print design for hundreds of years are equally valid online; in some cases, they are even more so'

Black (1997)

Due to the fact that WAP sites use primitive monochrome displays a thorough analysis of usability issues is required when considering developing a WAP site as mistakes in the design of the WAP site could make it totally unusable. This is due to the fact that unlike traditional web design there is no room to explain terminology through the use of rollover effects, icons or captions. Usability is concerned with determining the effects of interface design on the user with the goal of creating a usable design.

In order to clearly illustrate usability considerations when developing a WAP site the author has decided to create a WAP site for a fictional college named the **Sunshine Institute of Higher Education** or **SIHE** for short.

A number of frameworks were considered as a tool for analysing the usability issues involved in the creation of our example WAP site. On analysing these frameworks it was decided that a custom framework would best suit the authors needs, as there are currently no frameworks that are specifically aimed at WAP design, due to the fact that it is in its infancy. In developing the framework, background usability research into WAP design was carried out and a number of design methodologies were also looked at. The author has decided to use rapid prototyping in conjunction with user testing as a methodology for developing our example WAP site

3.1.1   Audience analysis

The end users of our example WAP site will be prospective and current IT students, therefore it is expected that end users will have a good knowledge of how to access the WAP. In order to ensure that the WAP site is appropriate for its designated end users individuals matching the end user criteria will be used at the testing and feedback stage. It is important to find test candidates who are representative of your intended audience; the creation of user personas can assist with this.

3.1.2   - Mobile platform analysis

The final product has been designed to run on WAP enabled mobile phones that conform to the WAP specification as defined by The WAP Forum. Platforms that will support our WAP site include:

- Nokia 7110
- Nokia 62210
- Motorola Timeport
- Ericsson R380s
- Ericsson R320
- Siemens c35
- Siemens m35
- Siemens s35
- Motorola a6188
- Motorola p73 89

3.1.3 – Development platform analysis

In order to emulate accurately the process of an inexperienced user building a WAP site I have decided to select an off-the-shelf WAP development package rather than coding the site by hand.   There are a number of WAP site development packages that could be used to develop the site. In order to decide which package would best meet my needs I decided to use a weighted rankings by levels table. To do this I first chose six areas that I felt were most important when using a development package. These six areas were:

1. Well structured GUI (Graphical User Interface).
2. Availability of a preview feature.
3. Design features of development package.
4. Extra features of development package.
5. Built-in WML Code validator.
6. Help system and availability of resources.

What follows is a discussion of each evaluation area.

3.1.4 – Well structured GUI

A well-structured development site GUI is important in the development of a WAP site. A poorly structured GUI would increase the time taken to use and learn the package and this in turn would increase the development time of the WAP site. Time management is essential when developing a piece of software thus poorly designed development packages must be avoided.

3.1.5 – Availability of a preview feature

The availability of a preview feature is vital when developing a WAP site. A real time preview window allows the developer to see what they are building as they progress. Development packages that make you save changes and load a card up each and every time you wish to view your progress can cost valuable time.

3.1.6 – Design features of the development package

The development package would invariably require a number of design features in order to build a high quality WAP site. Design features would ideally include:

1. Facility for creation of Wireless Bitmap Images (WBMPs).
2. Card title/Id display.
3. Facility to add an extra card to the WAP deck.
4. Facility to implement WBMP alt tag (will be shown when WAP device does not support WBMP images).

3.1.7 – Extra features of the development package

Extra features of a WAP site development package that would make it more desirable would include:

1. Facility to publish WAP site directly from the development package to a web server (integrated FTP client).
2. WBMP image editor/converter.

3.1.8   Built-in WML code validator

A built-in WML code validator is of great importance when considering a WAP site development package. Using a built-in code validator allows you to check your code incrementally as you build the WAP site, this allows errors to be identified and corrected immediately, as opposed to coding the site 'blind' and then coming across errors when you attempt to access the WAP site.

Table 2 lists the WAP development packages that were evaluated as potential choices and provides a short description of each.

Package	Description
WmlExpress 1.0	This program creates homepages for wireless Internet

	devices in just a few minutes. No WML programming skills are required. Just type in the title of the page, some text, and your WAP homepage is ready to be uploaded.
i-WAP WAP Site Generator 1.5	i-WAP WAP Site Generator 1.5 generates the whole WAP site according to its directory/file tree structure. So, one can build up or update a WAP site as easy as using "Windows Explorer"!
CoffeeCup Wireless Web Builder 2.0	CoffeeCup Wireless Web Builder is great for quickly building Web pages and Web sites for viewing on cell phones, PDAs, and other Internet devices. Get everything you need for creating and maintaining your Wireless Web site, all in one easy-to-use package. It has a powerful drag-and-drop site wizard, built-in live preview/emulator, a cool WBMP image editor with 100 ready-to-use images, a code validator and a site upload wizard
WAPtor 2.3	The WAPtor program is a simple but powerful WML editor for Windows 95/98, Windows NT, and Windows 2000 systems. It simplifies WAP page development by supporting the insertion of WML tags, offering easy-to-read color distinction of WML tags in the code of a WAP page, and showing the approximate appearance of the designed page on the mobile phone display. The WAPtor program allows opening and editing of existing WML files, or the creation of new files from a simple template. It enables you to work with more than one file at a time. The text editor is enhanced with color syntax highlighting to give users the opportunity to navigate through WML code more easily, and to make changes more effectively.
WML Editor 3.2	WML Editor 3.2 includes a Browser for previewing documents, a WML-Element-Helper and a WMLScript-

Helper for easy and fast writing.

Table 2. WAP development packages

It is clear that CoffeeCup Wireless Web Builder 2.0 is the best WAP development package for the authors needs. What follows is a description of the chosen development package.

3.2 - Development and test software

3.2.1 - CoffeeCup Wireless Web Builder

There were a number of reasons that I chose to use CoffeeCup Wireless Web Builder 2.0. The software incorporates some powerful features that are very useful when building a WAP site.

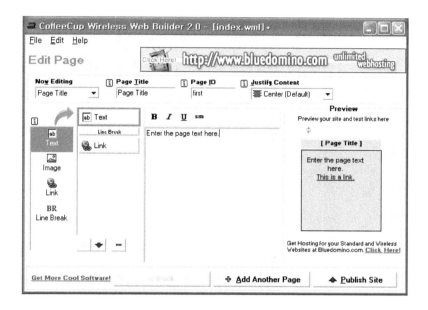

Useful features within CoffeCup Wireless Web Builder include:

Figure 4. CoffeeCup Wireless Web Builder

1. A built-in live preview feature which allows you to see pages as they are built.

2. A built-in code validator which allows you to ensure sure that pages will work correctly.

3. A Site Upload Feature, which allows you to easily publish pages.

4. A well structured GUI which aids the development process.

3.2.2 –WAP site emulation software

Effective testing of a WAP site as it was being built is key. Although the CoffeCup Wireless Web Builder has a preview feature, this does not necessarily mean that the WAP site will work across the majority of WAP phones. The only feasible way to test a WAP site as it is being built is through the use of various WAP emulators. A WAP emulator imitates the behaviour of a real WAP phone and is used by WAP developers to attain an idea of how a WAP site is likely to run on various phones and devices. What follows is an overview of the various emulators used when developing this WAP site in order to ensure effective functionality.

1. Official Ericsson R388s Emulator

The Ericsson R388s emulator (Figure 5.) is available at the official Ericsson website in the developers section and is provided as a tool for developers to test out their WAP sites on the new phone. The emulator uses the same BIOS as the real phone, which means it provides a 100% accurate emulation of the actual phone on a desktop PC.

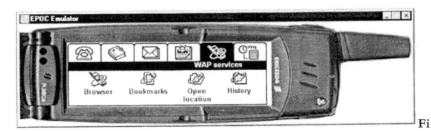

Fi

gure 5. Ericsson R388s emulator

2. Deck-It WML Previewer 1.2.3

Figure 6. Deck-It WML Previewr

The Deck-It WML Previewer 1.2.3 (Figure 6.) is also an excellent development emulator. The emulator by PyWeb software is based on the Nokia 7110 WAP phone, the previewer allows the author to get a good idea of how their WAP site will work on the 7110's display. The 7110 is one of the most, if not the most popular WAP phone currently available, thus it is extremely useful to test the site on a phone which a large majority of WAP users currently own.

In addition to emulating the features of the 7110 the emulator also has basic browser functionality through the buttons on the taskbar. The buttons allow you to go back to a previous page, reload the current page, stop loading the current page and also allow you to access your 'Favorites' in much the same way as Internet Explorer.

3. On-Line WAP emulators at GeIon.net

Gelon.net is a website dedicated to WAP technology. The site offers a variety of on-line emulators for various models of WAP phones. Although the emulation isn't 100% accurate, it still offers a valuable opportunity to gain an idea of how WAP sites would look on a broad range of WAP phones. At the time of writing, Gelon.net provided on-line emulators for the phones seen in Figure 7.

Fi

gure 7. Gelon.net emulators

3.2.3 – Interaction analysis

As the WAP site will be running on a mobile phone with a small monochrome screen rather than a fully-fledged computer and web browser, a number of factors must be taken into consideration:

1. Limitations to wireless devices. Information display size and limited input flexibility (such as a phone key pad) reduce usability of phone-based applications.

2. The limitations of wireless device processor capabilities, storage capacity, limited transmission speed.

3. Low network connectivity times due to battery limitations and channel occupation costs.

4. Time delays associated with information processing and access by means of the Internet.

5. The utility of the data being transmitted to the end user.

3.2.4 - Content analysis

There is a limited amount of media available to WAP phones. The two basic types of media that a WAP phone supports are text and images (wireless bitmaps), however not all WAP phones support wireless images, in which cases the site will be displayed as text only. The lack of available media types must be considered when developing a WAP site, image 'alt' tags must be used so that if an image is not available the user will be able to view text instead of a blank space where the image would normally appear. Images should also be used sparingly as the WAP device has limited memory and processing capabilities unlike desktop PCs.

3.2.5 - Usability research

Jakob Nielson, a recognised authority on website design and usability issues, has some invaluable design tips for developing WAP sites. A summary of his key WAP design points follows:

1. Avoid unclear labels and menu choices written in special language:

Users require simple language to be used in the design of WAP sites as there is no room to explain terminology through the use of rollover effects, icons or captions.

2. Avoid a mismatch between information architecture and the users' tasks.

Listing items in a mismatched fashion can cause frustration to the user. For example, TV listings that have a different screen for each channel would mean the user would have to wait for the WAP device to load each channels screen separately for the user to find out what was on all channels, at, say 8pm.

3.   Attempt to provide clear differentiation for your WAP site.

One WAP usability issue that has not been found on the web is a lack of clear differentiation between services.

'As one of our users noted when comparing the Financial Times and The Guardian: In the real world, you will have trouble finding two more different newspapers. On WAP, however, you can't tell them apart'

(Nielson, 2001)

4.   Employ precise task analysis

Very precise task analysis is necessary for WAP sites to be successful. As with the web, a great number of WAP sites suffer from poor task analysis, with many sites structured according to what company management thinks is best rather than how a user would typically approach the task.

Other academic sources provided usability guidelines useful for developing WAP sites. Detweiler & Omanson provided helpful general interface usability advice:

'No set of interface usability guidelines can answer the most important questions that must be answered before you design Web interfaces: What are the goals for building the site? What content will be presented? And, who is the primary audience? In general, your primary goals, e.g., to educate, to entertain, to sell, to foster creative expression, or some combination of these will dictate the content you provide and its unique look and feel.'

3.2.6 - Rapid prototyping – the spiral cycle

Rapid Prototyping, known as the 'spiral cycle' starts with concept definition and is then followed by implementation of a skeletal system (in our case an initial WAP site design) that is then followed by user evaluation and refinement (testing by users and modification of the WAP sit according to their feedback). This cycle then continues until the system is adequate. Rapid prototyping is most useful for the early development of a small-scale prototype used to test out key features of the design.

The rapid prototyping cycle can be seen below:

Rapid prototyping — the spiral cycle:

1. Concept definition.
2. Implementation of a skeletal system.
3. User evaluation and concept refinement.
4. Implementation of refined requirements.
5. User evaluation and concept refinement
6. Implementation of refined requirements in a continuous cycle.

Utilising a prototyping philosophy allows users the opportunity to feed back on the interface of the WAP site and its overall usability and to then try out the changes made, based on their original feedback, to see if the original issues are rectified.

3.2.7 - User testing and feedback

User testing and feedback will be through the use of questions and direct observation. As a rapid prototyping development methodology is being used, the testing will continue until the users are happy with the site. Three users who meet the criteria of being full time IT students in higher education will our test candidates.

CHAPTER 4

4.1 –Requirements specification

The starting point for the design of our WAP is a detailed requirements specification, an example can be seen below. This lists requirements at each stage of the design process and greatly aids the process of building a WAP site. The requirements specification was generated through surveying a sample of students in higher education through the use of questionnaires and then referring to the usability guidelines generated from the analysis of the answers. When creating a requirements specification you must also consider technological and design constraints associated with WAP in order to produce a feasible requirements specification that not only meets usability requirements but is also possible to achieve within the set technological and time constraints.

SAMPLE WAP SITE REQUIREMENTS SPECIFICATION

Sunshine Institute of Higher Education Requirements

1. Introduction

1.1   Purpose of this document

This document is provided as a set of requirements to guide the author of the WAP site.

1.2   Scope of this document

The WAP site is being produced as a deliverable for a book analysing WAP site usability.

1.3   Overview

The WAP site is to be a scaled down version of the full web based site. The WAP site will feature access to several key areas such as an on-line WAP prospectus, library, a news section and a contact section detailing lecturer contact details.

1.4   Business Context

This deliverable is not being sponsored by a business organisation.

2.   General Description

2.1 Product Functions

Access to the latest news at The Sunshine Institute of Higher Education, access to library information, a full prospectus and a Students Union area featuring a game written in WML.

2.2 Similar System Information

No similar system currently exists.

2.3 User Characteristics

The WAP site is aimed at both prospective and current IT students at The Sunshine Institute of Higher Education. This user group is expected to have a good level of expertise within the application domain.

2.4 User Problem Statement

Current and prospective students at The Sunshine Institute of Higher Education with WAP phones cannot currently access information with their phones, as a WAP site does not exist for The Sunshine Institute of Higher Education.

2.5 User Objectives Requirements:

Full prospectus, library Information, Students Union information, Lecturer contact details and Latest news.

Objectives:

Avoid unclear labels and menu choices written in special language. Avoid mismatches between information architecture and the users' tasks.

Attempt to provide clear differentiation for WAP site and employ precise task analysis.

2.6 General Constraints

WAP site must meet with the latest WAP specification on completion of the site, as released by The WAP Forum.

3. Functional Requirements

3.1 Description

Provide prospectus information, news from The Sunshine Institute of Higher Education, lecturer contact details and information on the SIHE Library and students union.

3.2 Criticality

All requirements are critical. Technical issues: Web space to be acquired from host and WAP site to be uploaded to server space via an FTP client.

3.3 Cost and schedule

No relative or absolute costs.

3.3.1 Risks

Ineffective time management and data redundancy. Avoidance measures including creating a schedule of work and regularly creating backups of files.

3.4 Interface Requirements

3.4.1 User Interfaces

The Sunshine Institute of Higher Education will be designed for access by the user with a WAP enabled mobile phone.

3.4.2 GUI

GUI will utilise text and images (Wireless Bitmaps).

3.4.3 CLI

No command-line interface is present.

3.4.4 API

No Application Programming Interface is present.

3.4.5 Hardware Interfaces

WAP gateway.

3.4.6 Communications Interfaces

HTTP 1.1

3.4.7 Performance Requirements

WAP site must function correctly within the constraints of a WAP mobile phone (low memory, low processing power, high latency).

3.4.8. Design Constraints

WAP and WML specification allow only text and images to be displayed (on a small screen). No other media types are available.

4.2 –Definition of WML decks and cards

A logical starting point when designing a WAP site is defining the various WML decks and cards that will be used within the site. As previously outlined within a WAP site a single file known as a 'deck' holds several pages, known as 'cards'. When the WAP device accesses a page the entire deck is loaded into the WAP phones memory. This means the phone can quickly peruse the pages or cards within the WAP site without having to keep loading each page individually. In our design several WML decks have been used due to the fact that if a single large deck is used the phone will attempt to download the entire deck into memory and generate an error message if it doesn't fit. Therefore splitting the site into several decks ensures that each deck will fit in the phones minuscule memory and that no errors will be generated, this is a sensible best practice to employ when you are building your WAP site. Figure 8 shows our site map, indicating the decks and cards that will be used within our site.

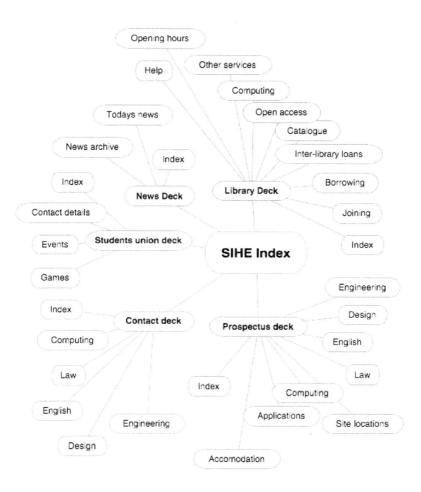

Figure 8. Decks and Cards Site Map

CHAPTER 5

5.1 - Implementation

Our sample WAP site was successfully built using CoffeeCup Wireless Web Builder, which was extremely straightforward due to its excellent user interface and live preview. As this book focuses on usability issues we will not focus on the specific build process within the development environment but rather will proceed onto the analysis and testing of our built prototype.

The prototype was tested by three IT students in higher education, the users evaluation was provided for each prototype and then modifications were made based on the feedback, the users would then test the modified interface and again provide feedback. This process would continue until all users felt that the design of the interface and usability of the WAP site was very good.

5.1.1 Initial prototype evaluation

The evaluation of the initial prototype revealed that users felt:

1. It was difficult to see where some text links began and others ended.

2. The Wireless Bitmap image at the top of each section within the site took up too much space and forced the user to scroll down the page to see the initial set of links.

This initial evaluation allowed the author to identify points of concern that were missed during the development of the initial prototype. The development of a second updated prototype will allow users to re-evaluate the site and to state whether they felt the interface was now acceptable.

5.1.2 Evaluation of updated prototype

Initial user evaluations were taken on board from the previous stage and the author created and then implemented a solution to the two problems outlined on evaluation of the initial prototype. The solutions were:

1. The solution to the problem of difficulty seeing where some links began and others ended was overcome through the use of the asterisk character ' * ' being used as a bullet point for each menu item, in much the same way as the bullet point in a word processor (Figure 9).

Figure 9. Solution to HCI problem of jumbled menu items

Problem: User evaluation of Initial prototype revealed that text links could become confusing and jumbled when there were a large number of menu items.

Solution: Second prototype solved this problem through the use of the asterix ' * ' character as a bullet point and sensible spacing.

2. The solution to the problem of the logo at the top of each page taking up too much space was solved through the resizing of menu items (Figure 10).

Figure 10. Solution to problem of Logo taking up too much space

Problem: User evaluation of Initial prototype revealed that page logos were taking up too much space and enforcing unnecessary scrolling.

Solution: Image is resized (made smaller), this allows more links to be visible and cuts down on unneccasary scrolling.

User evaluation of the second prototype revealed that the changes made to the initial prototype improved it greatly. Due to the modifications made to the WAP site users felt that the site was easier to navigate, which allowed them to access the information they required more rapidly. Users noted that the performance of the WAP site was also improved, due to the fact that the images that had been resized within the site (Figure 10.) now took less time to load. Users evaluating the site felt that it was now at an acceptable standard and that no further modifications were required.

5.1.3 Final interface adjustments

Some final adjustments were made to the prototype to ensure maximum usability:

1. The inclusion of a title at the top of each WAP card.
2. A 'back to the main page' link at the bottom of each WAP card (page).
3. The listing of the Sunshine Institute of Higher Education web URL on the main page in the format 'Visit our website: http:/ /www.sihe.com'.

The final WAP site interface design is now complete. The site retained the usability principals identified in Chapter 3, however it now also benefited from both further user and author analysis and testing.

5.2 – Implementation of media

The only media types other than text supported by WAP sites are wireless bitmaps (WBMPs). In order to create a wireless bitmap a conversion program must be used or the wireless bitmap must be drawn with a suitable package. The wireless bitmaps on the Sunshine Institute of Higher Education site were created using the image conversion utility bundled with CoffeeCup Wireless Web Builder known as CoffeeCup Wireless Image Maker as seen in Figure 11.

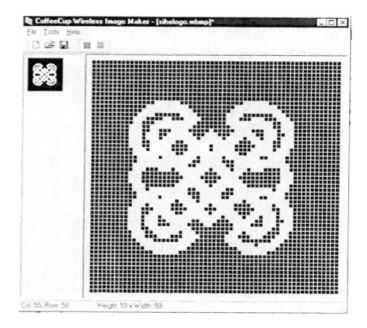

Figure 11: CoffeeCup Wireless Image Maker

5.3 – Implementation conclusions

Using rapid prototyping to refine the tested prototypes based on user feedback allowed updates to be made quickly and tested for effectiveness. The three main mechanisms utilised within this project to ensure a high level of usability were:

1. Usability analysis findings.
2. User testing of WAP site and interface.
3. Author's evaluation of interface.
4. Iteration of interface changes and re-evaluation by test subjects.

The development environment - CoffeeCup Wireless Web Builder allowed easy and quick updates to the text, images and links within the WAP site in an efficient manner by displaying the site as a set of components as seen in Figure 12.

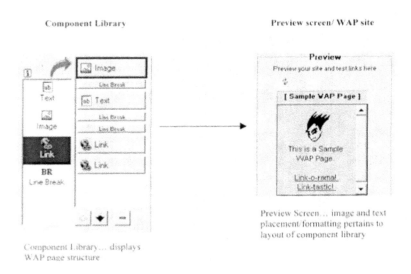

Component Library.... displays WAP page structure

Preview Screen... image and text placement/formatting pertains to layout of component library

Figure 12. Wireless Web Builder component structure

This segregation of the WAP card or page allowed various sections of the WAP page to be edited quickly and efficiently, whilst allowing all pages to be previewed as they were edited.

The format of the development tool was ideal for a rapid prototyping approach due to the ease and speed of modifications.

During the implementation phase of this project, in fact throughout the entire project, interface and usability issues have had an extremely high priority. When a WAP site is viewed on a wireless device there is very little information available aside from the text and links that you place on the WAP card (page), meaning that a poorly designed interface could lead to utter confusion for an end user, therefore when designing and building your WAP site you should pay a great deal of attention to these issues.

CHAPTER 6

6.1 - Results

This section describes the finished WAP site and evaluates its effectiveness.

6.1.1 – Annotated screen shots

Annotated screenshots from the finished WAP site can be seen in Figure 13

Figure 13. WAP site

Screen: Main welcome screen

Purpose: Provides access to the rest of the WAP site.

Screen: Library

Purpose: Provides information

Screen: Prospectus

Purpose: Provides complete on line version of the 2001 prospectus.

Screen: Contact Details

Screen: Students Union

Screen: News

6.1.2 – Description of finished product

The finished site functions correctly and due to the analysis of usability factors presents a clear and logically structured menu system and intuitive interface. The WAP site was tested on a number of different WAP phone emulators in order to ensure that it functioned correctly.

6.1.3 – Testing of WAP site

Functionality Analysis : WAP Emulator Tests

In order to ensure that the WAP site functioned correctly on a variety of WAP phones, utilising the on-line WAP emulators at Gelon.net as shown in Figure 14.

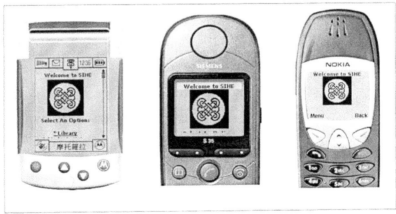

- User Testing : End User Tests

Figure 14. Screenshots of SIHE WAP site being tested on a variety of WAP phone emulators

CHAPTER 7

7.1- The future of WAP

7.1.1 - New methods of doing business

M-Commerce, a new method of conducting business and delivering services that WAP technology allows to be possible, was covered by the author in chapter two. Research showed that for M-Commerce to be successful the currently inadequate mobile infrastructure must be adapted or changed altogether.

7.1.2 - The business benefit of WAP

The new methods of conducting business that WAP technology allows do have some benefit. Examples of useful WAP sites such as Lastminute.com, Yahoo!, Excite and The Guardian WAP show real business value.

7.1.3 - Current mobile infrastructure

The current mobile infrastructure (GSM Mobile Network) is not ready for the wide scale adoption of WAP in the UK. Two superior networks namely UMTS and GPRS would make M-Commerce more of a widespread viable option.

7.1.4 - Is WAP a fad?

From the research carried out in this document the author can safely say that WAP is not a fad, but it may well be superseded as network speeds improve and more powerful devices are released. WAP has industry wide support and a number of big players are backing the technology including Nokia, Ericsson and Vodafone, however competing technologies such as HDML (Handheld Device

Markup Language) and i-mode are already available in Japan and offer superior services, operating on their Personal Digital Cellular Packet (PDC-P) network. Whichever technology becomes standard, mobile Internet access is set to really take off in the coming years.

CHAPTER 8

8.1 Overview: WAP Case Studies

This book has demonstrated the importance of usability considerations when designing and building a WAP site. This section provides case studies of WAP and WML being utilised in the real world. Each case study looks at the effectiveness of WAP within a particular hotel and outlines the pros and cons of the technology in practice. All hotels surveyed were early adopters of WAP technology and case study data was obtained from a working draft prepared by the Centre for Regional and Tourism Research in Denmark.

8.2 Listing of WAP case studies

8.3 - Scandic Hotels

8.3.1 - The firm

In terms of revenue, Scandic Hotels is the largest hotel brand both in Sweden and overall in the Nordic region. As many as 133 of its 155 hotels are located in the Nordic region, i.e. 86% of its hotels, and this region (Sweden, Norway, Finland, Denmark) also accounted for 86% of Scandic's total revenue of 707 million Euros (6 billion SEK) in year 2000 (annual report). Sweden alone accounted for 52% of Scandic's revenue. Scandic had 72 hotels in Sweden by the end of year 2000, with an average of 182 rooms per hotel. Scandic accounted for 15.6% of all the hotel rooms in Sweden in year 2000. Scandic's room capacity utilisation was 63%, i.e. higher that the general Swedish average of 48%. Scandic Hotels employ approximately 7,500 people. In the middle of year 2001 Hilton Group plc acquired Scandic Hotels AB. The plan is to convert a third of the Scandic Hotels into Hilton.

8.3.2 - The Website of Scandic Hotels - and WLANs

On the Website of Scandic Hotels it has been possible to make online real-time bookings since 1998. In the year 2000 the number of fixed-line Internet bookings were still modest, but increasing quickly.

Back in March 2000, Scandic Hotels announced a WLAN initiative in cooperation with Wirelessbolaget/InfraLAN and Nokia. Wirelessbolaget listed just two WLAN locations in Sweden on their Website by the middle of 2002, one of which is a Scandic Hotel - the Jarva Krog in Solna, Stockholm, which was also the single Scandic Hotel in Sweden mentioned in the March 2000 announcement. Jarva Krog does mention wireless Internet as one of their conference facilities.

8.3.3 - The WAP site of Scandic Hotels

In January 2000 Scandic Hotels became "the first hotel chain in the world to offer their guests the opportunity to access services via a WAP-enabled phone." In the first instance the service was offered to their most frequent Swedish guests - the so-called 'gold' members of their loyalty programme, i.e. those who stay at least 60 nights per year.

Four items were included in the menu of the WAP solution of Scandic Hotels:

1. Booking of a hotel room at all Scandic Hotels within the Nordic regions. The booking was confirmed directly by a personal SMS message. Cancellations, changes and checking of room availability could also be undertaken.
2. Under the account balance menu item, the Scandic Club members could see how many free room nights they had in their account.

3. Under the pleasure weekend item, the WAP-using Scandic guest could obtain information about events, concerts and theatre plays.

4. Links to several external information sources were provided: News from the national newspaper Aftonbladet.se; weather forecasts and news from the Swedish Stock Exchange, as well as currency exchange rates from Forex.

Nocom AB10 was responsible for project planning, deployment and software, Nokia for phones and WAP server technology, and Geab - The Phone House for packaging and delivery to customers. Adera undertook the WAP application development and Europolitan (Vodafone Sweden) was involved as well. In October 2000 Nocom and Europolitan established the jointly owned company Mobile Relations AB (MR) for the marketing and sales of mobile Internet for businesses, including mobile Intranet and Extranet solutions. MR, which has five employees, has announced several SMS-based solutions but none based on WAP. MR's revenue was 7 MSEK in 2001, i.e. a little under 1 million Euros, but the net result remained negative.

On 1 April 2002 Scandic Hotels closed its WAP site because there were simply too few users.

8.4 - Tiscover AG Travel Information Systems

8.4.1 - The firm

TIS was founded in 1991 as a subsidiary of Tyrol Werbung. Initially a Destination Management System (DMS) for Tyrol was developed, then Corrintia. TIS originally meant Tyrol Information System, but later took on the more general meaning Travel Information Systems. According to its own statement, Tiscover is now "Europe's leading provider of Online Destination Management".

8.4.2 - The Website

The first generation of TIS on the Web was opened in 1995. Since the second online version was established in 1996, the name of TIS on the web has been Tiscover. With Tiscover 2002, the site is in its 7th generation.

Tiscover has become a very well-known DMS, with sites in Austria, Germany and Switzerland. In Germany Tiscover cooperated with Start Media Plus. The site with content covering Germany was originally called Deutschlandreise.de, but is now called Tiscover.de. Similarly the Swiss version is called Tiscover.ch. The Austrian version, Tiscover.at, remains the most important of the three sites. From the end of February 2001 a uniform layout for all three sites was introduced, which is a benefit for the users. Among travel-sites visited by Germans, Tiscover.com have been ranked in the top 10 repeatedly by the Internet research firm Jupiter MMXI.

The Tiscover site generated over 200 million Euros worth of bookings in 2001, excluding telephone or other off-line bookings inspired by information from the Web. This probably made Tiscover one of the top five travel and tourism related Websites in the European market in 2001 in terms of online booking revenue generated.

A lot of effort has been put into improving the bookings system. It works on an allotment base, but a Tiscover online booking is a guaranteed booking. Bookings can be guaranteed with credit card payment or by bank transfer.

The Website visitaustria.com leads to a strictly Austrian version of the Tiscover site, and so does the address Tiscover.at, whereas the top-level address Tiscover.com leads to a common entry point for Austria, Germany and Switzerland, plus Liechtenstein. Visitaustria.com is built on Tiscover but is the domain name of the national Austrian tourist board in Vienna.

8.4.3 - The WAP site

Late in 1999 the decision was made to develop a WAP version of Tiscover.at. The WAP site was launched January 2000. This was simply the earliest possible launch date.

In the development phase important elements were defined. Highly variable types of information were to be included: weather situation and forecast, snow situation, avalanche risks and warnings, ice situation on lakes (important in Corentia), temperature of lakes (summer) and events as well as accommodation information. Only available accommodation (hotels and similar) is shown at the Tiscover WAP site, which has the address http://wap.Tiscover.com.

The WAP-pages are created dynamically, i.e. as and when requested by the WAP-user, in a process controlled by Java servlets.

Initially, the Tiscover WAP development team had no WAP phone, therefore the WAP site was built with the Nokia development kit. In the launch month, January 2000, they received a Nokia 7110. Since the screen of a WAP-phone is small, design and usability considerations are especially important. Screen sizes vary between phone models, which is one of the problems when designing WAP sites. For example, the Nokia 7110 has a bigger screen than the early WAP-phones from Siemens and Motorola. Furthermore, there are different WAP-browsers, between which the behaviour of select boxes differs. It is therefore necessary to test WAP sites on different phones and browsers. The Nokia 7110 was the first and the most important WAP-phone from a WAP-developers point of view accounting for the greatest number of user sessions on the Tiscover WAP site. The Tiscover team also tested the WAP site on several other WAP-phones.

At the entry page of the Tiscover.at fixed-line website, in both the German and English version, there is an icon with a hyperlink to information about Tiscover on WAP. The icon has been there since the WAP site was introduced in the beginning of 2001, and is still there in the middle of 2002. Given the great number of fixed-line website visitors, even with this single initiative, the WAP site of Tiscover has always had and still has a quite prominent position in the overall market communication picture of the Tiscover company.

During 2001 Tiscover counted about 75 WAP-user sessions per day, i.e. about 27000 in total. This was dwarfed by fixed-line usage, though. Each time there was one WAP-user session there were 1950 visits to the Tiscover fixed-line website, i.e. a ratio of almost 1:2000 in favour of fixed-line. Initially there were only a few WAP-phones in use. And slow connections were one of the factors that deterred users. This will improve with GPRS, mobile GPRS networks are now in commercial use all across Europe and by the four mobile network operators in Austria. GPRS is expected to have a positive impact on WAP uptake. This, however, also requires GPRS handsets of course.

By the end of 2001, only 3.3 million GPRS phones were in use in Western Europe, and "less than one third of the device owners were using them to access packet-based cellular data services". In other words, there were no more than one million GPRS users by the end of 2001, out of about 291 million registered mobile phones in a population approaching 400 million, up to 30 million GPRS phones may be sold during 2002 however. There could therefore be up to 33 million GPRS phones in use in Western Europe by year-end 2002. Up to 30% may use GPRS, i.e. there could be up to 10 million GPRS users by year-end 2002, but most likely less, say 8 million, or no more than 2% of the population. This could increase to 4 or 5% in 2003, though. - For WAP in general, i.e. either over GSM or GPRS, 28 million phones could be used for WAP by year-end 2002, by 25 million unique (different) peoples i.e. about 6% of the population (up from 4% in 2001). WAP users may increase to 9% of the population by year-end 2003, whereas fixed-line web-users could reach about 45% of the population in Western Europe.

Mobile phone density in Austria was as high as 74% and 80% by the end of 2000 and 2001, respectively, well over the European averages of 62 and 74%. The population in Austria is a little over 8.2 million. There were around 400,000 WAP phones in use in Austria by the end 2000, possibly up to 500,000. Around 200,000 WAP phones were used for WAP by the end of 2000. This number had more than doubled to about 430,000 by year-end 2001. This was 3.3% and 6.5% of the installed base of mobile phones in Austria by the end of the years 2000 and 2001. These percentages were above the estimated European averages of 2.9 and 6.3% respectively. The Tiscover team even believe that the vast majority of WAP users (up to 9 out of 10) only have used WAP a few times, leaving only few as frequent WAP users.

The Tiscover WAP-team has built two different accommodation search functions: One for business travellers and one for tourists. The difference is in the search possibilities; the business travellers wish to search for a hotel in a specific town or city quickly, whilst tourists have quite different requirements.

The menu items on the site are:

- Weather report

- Snow report

- Ice report

- Events

- Room booking (search) - Leisure

- Room booking (search) - Business

- Information about Tiscover WAP

Currently it is not possible to make online bookings on the Tiscover WAP site, but this feature may be added in the future, therefore, Tiscover currently only utilise WAP to provide information. However the user can call to make the booking by voice, by clicking on the dial-up link on the WAP site. Making the booking via voice is considered more user friendly than to make the booking via WAP. Tiscover track the number of WAP-user sessions but do not currently count the click-throughs via WAP to individual properties or overall but if the WAP-application was built in a certain way it would be possible.

The Tiscover WAP developer team members are quite happy with the WAP site although they see possibilities for improvement. In December 2000 location sensitive services (or: location based services, LBS) became a possibility in Austria. This is an interesting opportunity for travel and tourism related WAP sites in general and for Tiscover specifically. With LBS information can be personalised, taking current location into account. Tiscover hope to introduce LBS in 2002.

In addition to the end-user site on WAP, Tiscover utilises a couple of additional mobile services:

1. Extranet: Hotels and other accommodations can control their allotments via SMS, i.e. they can send and receive vacant/occupied rooms reports via SMS.
2. Mobile payment: Thanks for a strategic partnership formed between Paybox Austria AG, Tiscover and the Austrian hotel association, it is now possible to pay hotel bills by mobile phone, at least at some hotels in Austria. The Paybox system does not build on WAP, though. In stead the voice-function of the mobile phone in utilised. Keying-in of pin-code is required, and the amount is drawn on the bank account of registered users. Paybox is currently not one of the payment options at Tiscover.com.

The number of mobile phones and other mobile devices will increase in the future. GPRS will arrive as will UMTS. This will lead to greater user-friendliness, be it via WAP, i-mode or another technology such as XML.

8.5 City-guides.ch

8.5.1 The firm

BeKoBasel (short for Beratung und Kommunikation, Basel) was founded in the year 2000. The firm is run by its owner, Ms. Ursula Rhein, but when needed, freelancers are used. Ms Rhein worked as an architect until WAP arrived, but then decided that she could not do both. The firm offers Web and in particular WAP advice and solutions. The firm does have a website with basic information about the services offered, but the firm's primary focus is on its WAP site. On the website there is an emulator, which shows the content of the WAP site, i.e. six Swiss City Guides and more.

8.5.2 City-guides.ch on WAP

The Basel-guide was the first City Guide on WAP, which BeKoBasel developed. The five others are: Bern, Geneva, Lausanne, Luzern, Zurich. All the six city guides can be accessed via a portal called city-guides.ch. Also Messe Schweiz is in the portal, including a list of major ongoing or upcoming fairs in Switzerland. Finally the portal contains links to useful information such as: Weather (by wetteronline.de), Avalanche Bulletin, SBB Timetable, Swissinfo , currency converter, WAP translator (German-English and visa versa), and contact details - for sending comments to BeKoBasel. Each guide comes in a German and an English version, some even in French as well.

The city guides are on the 3 Swiss mobile operator portals: Swisscom, sunrise (diAx) , and Orange.

Hotels are listed with address, phone number plus e-mail and Website address, if available, and map (for the particular hotel).

In the main menu of the city-guides.ch portal there is a list of all city guides in portal plus links to several other WAP site with very useful information for residents and travellers alike, and both groups visit the site. BeKoBasel works with the region's TV station to get content. She puts their news on the WAP site. There is a focus on culture, which makes the WAP site interesting also for local people. Thus, many from Basel visit the Basel-guide on WAP.

The WAP site was established by the end of February 2000. There were about 2400 visits per month during the six months to and including January 2001, and 2.7 page views per visit in average during the same period.

Hotels pay to get listed in the WAP guides. However, BeKoBasel only asks hotels that they know and rate highly, to join the system. The WAP city guide is like a printed city guide. It contains ideas about where to go, what to do, and unlike printed guides it lists fairs and conventions. Culture is the most visited part of the WAP city guide: Museums, theatres and cinemas as well as news, restaurants, and hotels, people from abroad are interested in the latter. Contact details for consulates and emergencies are also listed and the Swiss City Guides are for business and leisure travellers alike. BeKoBasel also works with Messe Schweiz, which is based in Basel. From their survey BeKoBasel knows that an important user segment is in the age group 25-50, i.e. young to middle age business people,

BeKoBasel utilised the Nokia toolkit to develop the WAP site. The WAP-pages were busy however and BeKoBasel had to be careful about size and letters/signs. The actual site was tested on different WAP phones: Nokia 7110, Ericsson R320s, and a Motorola model.

In spite of outsourcing the mobile phone production as announced early in year 2001, Ericsson, in association with Swisscom, was very active in the promotion of WAP content development, and launched its Mobile Applications Initiative (MAI), also in Switzerland, to support application developers.

BeKoBasel is optimistic about GPRS. Several Swiss mobile network operators (Swisscom and sunrise/diAx) started GPRS tests in 2000, at a time when only the Motorola GPRS phone was available, the Accompli 008. By the beginning of 2001, WAP had got a rather bad reputation in Swiss newspapers however with GPRS access speeds should be much faster. Swisscom's commercial rollout of GPRS started on 1 February 2002. With GPRS, transmission speeds of up to 30-40 kbit/s are now possible and it is therefore ideal for WAP services. BeKoBasel got an early GPRS test phone from Ericsson, the R520. Other GPRS phones are to follow.

Swisscom possibly had between 150,000 and 200,000 WAP users by the end of 2000. The Swiss City Guides count about 100 visits per day. In 2001 it was thought that 50% of WAP phone owners in Switzerland actually used WAP about once a week. Up to 70% might use WAP at least once a month.

The next big step in WAP for hotels will be direct booking. Currently WAP is not secure enough, There is no security. City-guides.ch will not embark on adding booking via WAP until at least the same level of security as on fixed line Internet is available on WAP. City-guides.ch does have a click-through function on its WAP site, to enable users to get through to make a booking by voice. The voice function of a mobile phone is adequate for most people. To send credit card details on a WAP phone is not only cumbersome, but also insecure. Therefore voice is the best way to make a booking on a WAP enabled mobile phone at the moment.

Hotels like e-commerce. Hoteliers would like to be able to offer the same on WAP as they do via the full Web, they say. We do not really know if customers will make use of a WAP booking function however if it was offered. Hotel booking services may be offered which do not require the traveller to state name, address and credit card details each time. This is based on user registrations on fixed-line Websites.

A WAP site has to be carefully designed specifically to suit a small screen on a WAP phone, otherwise it would have poor usability. This would also be the case if WAP browsers were able to read ordinary web pages. BeKoBasel really believes in the future of city guides on WAP, and might even expand outside of Switzerland.

8.6 2PL

8.6.1 The firm

2PL is owned and run by Mr. Peter Lobel. The 2PL name was inspired by the fact that Mr. Lobel's wife's first name is Petra. Peter Lobel does most of the work himself, but also makes use of about 15 freelancers. 2PL is thus a virtual firm. Mr. Lobel has previously run a real firm, but will not try that again as he enjoys his current lifestyle, on the French Riviera too much.

8.6.2 The Website

2PL.com is a multi-country, multi-language hotel-website that was started back in 1995. It currently contains information about 6000 hotels. The same content is shown in as many as 18 different languages. This is done as for the majority of people English is not their first language. Holidays are important for people and

Mr. Lobel made the decisions that people should view information in their own native language. In the future people will have more leisure time and Mr. Lobel does not see any big difference between business and leisure travel, however more information is needed for leisure. The plan is to extend the Website with more cities and more hotels. The 2PL.com Website counts about 1500 visits per day, max. 2000. There was a great increase in visits to the fixed-line Website from 1999 to 2000 (300%), and a further increase from 2000 to 2001. 2PL puts the hotels in the system initially and at which point the hotels (should) maintain the information themselves. Mr. Lobel believes in the idea of community building and has a community discussion board on the site. 2PL can see when a booking enquiry is being made.

2PL does not make much money from the Web-info. It is difficult to develop a viable business model: Too many Internet service providers want money from the hotels. Rather than start by going to the hotels and asking for money, it's a better starting point to offer something of value to the hotels such as common purchasing, which can help hotels make savings. There may be savings to be made from common purchasing: 20%, of which 5% goes to 2PL. 2PL normally goes for the small hotels, although the big hotels are also welcome, there are also B&Bs and pension houses listed in the system. A customers confidence must be won before you can do business with them.

8.6.3   The WAP site

Content pages - for Web and WAP - are written in XML. 2PL started developing WAP in December 1999, and went live in February 2000. Mr. Lobel uses a Sony Z5 which contains an HTML browser. Not only does the 2PL site come in a WAP version but it was also developed in an i-mode version from the outset, i.e. a version written in CHMTL, compact html. I-mode is considered to be faster and cheaper than WAP. From the outset i-mode charged content per data-package, whereas WAP was and generally still is being charged on a time basis, for example 0.20 Euros per minute. Problems with WAP are that it is too expensive, there is too little information, and it's too difficult to navigate on WAP. For information that is needed immediately, WAP is OK. However WAP is not an appropriate medium for information about next week. WAP should be used for immediate information that is needed here and now, 2PL offers its hotels information in a WAP version but they do not make much money from it currently.

It only took a few days of work to develop the WAP site but there was no standard upon which to base the development. The WAP site was tested with the Nokia developer kit. As a database, 2PL uses SQL from Microsoft. From the outset there was no click-through from the hotel phone number in the WAP site, but 2PL may enable that function.

Normally there are 1000 hits on the 2PL WAP site per day, corresponding to 50 visits on the WAP site per day at its maximum, with 20 hits per visit or each user session. The perspective on WAP for hotels? It's good for the initial contact details, then potential guests can phone the hotel and make a booking enquiry and the booking by voice.

The WAP site of 2PL is no longer live.

8.7 Hotelguide.com

8.7.1 The firm

Hotelguide was founded in 1991 as a publishing company, issuing a printed international hotel directory. This was later supplemented with a CD-ROM version. Today Hotelguide has 35 employees, mostly sales people, of which 25 are based at the corporate headquarters, recently moved to new premises.

8.7.2   The Website

In 1995 the first Web-version was launched. Currently the Hotelguide.com Website contains 65,000 hotels in five - or eight - languages. Hotelguide have invested in internalising all IT functions, such as webhosting. The current version of the system went live in March 2001. It runs on an Oracle database.

There are at least 360,000 hotels worldwide, of which only about 60,000 hotels are in GDS' such as Amadeus and Galileo. Hotelguide.com will soon move to real-time pricing. Hotelguide.com ensures that the hotel will be distributed through different channels to the end consumer.

Hotelguide.com counted 130-150k page views per day in 2001 (140k*30=4.2 million per month), corresponding to 20k-28k visits per day (25*30=750k per month), possibly representing 12k-15k unique visitors per day. There was a 30% increase in Website traffic from 1999 to 2000.

Hotelguide.com makes multiple availability checks. There are 25k hotels in Amadeus alone, with availability information. Hotelguide.com gets about 100 online-bookings per day, with commission. On top of these come booking requests made directly to the hotels, without commission. For those hotels with online availability information replies to reservations requests are of course instant.

Hotels pay to stand out from the masses. This will never change, irrespective of technology. Those hotels that pay get listed first on Hotelguide.com search results.

The commission today is 8-10%. It will drop however, perhaps to 5%. Those who base their business model solely on commission will therefore be in trouble. In the hotel intermediary business there are frequent mergers and closures, perhaps there will be possibilities in the future to charge for services like mobile

Internet (from users). Mobile Internet access via WAP is an opportunity for last minute hotel bookings.

8.7.3   The WAP site

Hotelguide started development of a WAP site in December 1999. It was launched in January 2000. An i-mode version was launched in April 2000. Both included basic information and functions. Next steps will be availability checks and booking on WAP. There is no particular target group for the WAP service, but the first users are expected to be business people. There is an automatic click-through function on WAP. Nokia, Ericsson, and phone.com micro browsers enable click through from the WAP site to get through to the hotel on the phone without manual dialling. Today hoteliers accept telephone bookings without any credit card guarantee.

If travelling to another time-zone, non-voice booking on the WAP phone is preferred. This may be combined with guaranteeing the booking by credit card. For non-registered users this would require the keying in of 16 digits. For those already registered it would be easier. In the hotel business the credit card number is just used as a guarantee, if used at all. The actual payment is settled at the hotel. There is little or no safety on WAP. If someone wants to misuse credit card details it is possible.

On Hotelguide.com guaranteed bookings on WAP are not possible currently. Hotelguide.com has to get partners involved to implement this. Payment can also be via partners, everything on WAP has to be user friendly. With regards to acceptance and security: time will tell. Initially Hotelguide has made a basic WAP version available.

The Hotelguide content is used by Shell Geostar on iPAC. Using this device there is a bundling of information: Driving directions, weather forecasts and hotel accommodation. Hotelguide.com expects to be involved in a lot of WAP OEM-projects, i.e. in the business-to-business field, bundling hotel information with other types of information.

The WAP site was developed in-house, using a trial and error approach, with lots of enthusiasm. As a basic WAP site it works fine but it is problematic that text displays differently on each device: Nokia, Ericsson, ... etc.

The long-term plan is to enable checking of availability via WAP and to enable the display of maps with location and driving directions to hotels. They are already all geo-coded and Hotelguide have positive feedback on their WAP site, in fact it has won several awards.

The expectation was that the WAP site would be used as much as the Website, however, the reality is that there are 50 times as many visits to the Website than to the WAP site: There are about 25,000 visits per day to the Website vs. about 500 visits per day to the WAP site.

The WAP-version of Hotelguide.com site is on many WAP portals. This should provide many visitors. The idea is to share revenues:

- Operators share airtime or data transfer revenue earned from hotelguide.com's service.
- And Operators earn a fixed fee from hotelguide.com for each confirmed booking via GDS (Global Distribution System).

Hotels today get at least 80% of their bookings via telephone, as travellers like the real interaction with a person. Most new phones will be WAP enabled.

Location based services (LBS) will be an interesting opportunity, it is already available in Switzerland:

With location-based services, the location is determined via the mobile phone such that each time the phone is activated it establishes contact with the base station with the strongest reception. The smaller the cell of the base station, the more exact the location. In built-up areas the size of the cells is generally a few hundred metres while in rural areas they can be several kilometres. Swisscom Mobile has developed a platform specifically for location-based services, with the positioning data evaluated and prepared according to uniform criteria and within the framework of the Swiss Federal Law on Data Protection.

Other future buzz words will be: Co-opetition, net partnerships, intelligent services (including LBS). Co-opetition is the co-operation with competitors, i.e. service offers are partly overlapping. Net partnerships may include for example taxis and local (destination) information.

Hotelguide as a company remain very optimistic about WAP, in spite of the WAP flop that was seen by the end of 2000. New market stimulation will come from GPRS and 3G and Hotelguide.com expect WAP to become a success by or before the end of 2002!

Major mobile phone manufacturers promoted WAP heavily but only a low percentage of WAP phones are actually used to access WAP sites regularly. Some WAP phones are not even configured properly. WAP-phone owners often stop using WAP after just a few attempts, it appears that there is an equal split between those WAP-phone owners who have never used WAP (a third), those who have just used WAP one or twice (a third), and finally regular users and even infrequent users (a third). - All the content of Hotelguide.com for the different platforms is written in XML. This facilitates integration and provides flexibility.

WAP phones may start outnumbering the PC but WAP usage will not overtake PC usage anytime soon for Internet access either for hotel related information or for other types of information.

8.8 Book online, ITS, Italy

8.8.1   The firm

ITS (www.itsitaly.com) was started in 1997, in Padua, Italy. The firm has about 40 employees. The main area of activity is Web-solutions and ITS is now an IBM Certified E-Commerce Premier Business Partner. The products of ITS include:

- An end-user E-business software solution called BuyLand.

- The hotel-booking site www.bookonline.it.

8.8.2   Bookonline.it on the Web

By the end of September 2001 the bookonline.it website returned 9 * 98 = 882 hotels. For all hotels the name, town, telephone number and number of stars are mentioned at the top level followed by three options for each hotel:

1. "Book now": offers a list of costs of your stay with the possibility to reserve your hotel and pay on line.
2. "Ask for availability": permits to send an e-mail to the hotel to ask about availability, rates and further information.
3. Basic details only for off-line contact.

The Website includes the following features:

1. Those who register with name, address, phone number and e-mail address can make secure online bookings.
2. Registered users can log on with their chosen user name and password.
3. Hotels can be searched region, and then by city/town within each region.
4. Last Minute Offers.
5. It is possible to send a BookOnLine E-Card.

6. It is possible to buy guide-books online, in association with Amazon: 14 titles shown.

7. Useful links: There are 7 categories of useful links, 41 in total.

8. Finally there is a WAP emulator which enables Website visitors to see what the WAP site looks like.

The objective of developing the Bookonline.it website back in 1997 was to get experience within e-commerce, with hotel booking chosen as a good application.

After a six-month period, development of the online solution was finished in 1998. It helped ITS propose/introduce their E-commerce competencies into the market. However, the tourist market is not strategic for IBM. In 1999 BookOnline.it was introduced at the BIT 1999 fair/exhibition in Milan (in co-operation with IBM). BIT is the most important tourist fair in Italy (an Italian version of ITB in Berlin).

The Web-version of Bookonline.it has two parts:

1. A front office for the customers.

2. Back-office

It is possible to search for hotels by region and then to make either an online booking or to send a booking request. In order to make an online booking the customer has to be registered, overall the website is mainly visited by leisure travellers since bookings and booking requests are for holiday periods and weekends.

The hotel can update their availability data themselves in the back-office system. To get access they have to key-in a user name and password. - By the end of December 2000 there were 651 hotels in the database and by the end of September 2001 882 hotels.

Hotels are normally contacted by e-mail to gauge their interest in joining the system. To join the online booking part of the system hotels have to sign a contract with ITS. Only a small percentage of all the hotels that are listed in the system are bookable on a real-time basis (with availability data). Most bookings are on a request basis only. Hotels generally prefer to get the booking directly, however few customers are happy to provide their credit card details online.

There are many family hotels but few chain hotels in Bookonline.it. Many of the hotels do not have an electronic reservation system. Hotels want to be able to be flexible and they do not want to give allotments. Prices can vary, to influence demand. Some hotels will not take Internet payments for accounting reasons however; there is a trend towards payment online.

Bookonline.it count 250,000 to 300,000 hits per month and 100,000 page view per month, on their website. This results in about 1000 reservation requests per month.

8.8.3 Bookonline.it on WAP

By the end of 1999 ITS put together a group of three or four people to study the feasibility of building a WAP application, it was concluded that WAP was not ready for heavy use. Later a person from the initial group developed a WAP interface for BookOnline, which was deemed to be a good idea. It was not difficult to develop the WAP site and the WAP-version of Bookonline.it was launched in March 2000, as a simple listing of hotels with only basic information. A progression of the site would be online booking via WAP. Mobile phone penetration in Italy is very high, among the very highest in Europe and consequently WAP penetration is also quite high.

Bookonline.it on WAP includes 445 hotels, unchanged since May 2000, i.e. there are about half as many hotels in the WAP site as in the fixed-line Website. The number of hotels in the latter is being increased continually, whereas this is not the case for the number of hotels on the WAP site. ITS decided that WAP was not a big part of their strategy. ITS await a maturing of the market for WAP applications and newer network technology.

All hotels are displayed on WAP, i.e. not only those with available rooms. It would not be a technical problem for ITS to display only the hotels with available rooms on WAP, however this would require the user to key in the date of arrival, number of nights etc. and it would require the establishing and maintenance of a database. Since ITS has a web-booking site, requiring customers to register before they can make a booking on the Web, the same personal data could be used to allow registered customers to make WAP bookings. ITS has conducted a customer survey which shows that 40% would be willing to book on WAP.

Online booking on WAP could be implemented swiftly but the company have decided to wait for UMTS. WAP on 2G is currently for displaying information only.

8.9 Thistle Hotels, UK

8.9.1   Thistle Hotels - the firm

Thistle Hotels plc is one of the UK's leading four-star hotel companies with 56 properties (just under 11,000 bedrooms) in key city centre and countryside locations. 18 of the hotels are owned or leased and 38 are managed. With 22 hotels (about 6,000 bedrooms) in London alone, it is the capital's largest hotel operator.

Corporate guests account for close to 60% of room nights sold. - With international business travellers from the US and other oversees markets being an important customer group for Thistle Hotels, the Group was heavily affected by the negative impacts on travel markets of the tragic events on September 11, 2001.

8.9.2   The Website of Thistle Hotels

For years, Thistle Hotels has focused on reducing reservations costs by channelling more bookings through electronic channels, such as its own Website and the Global Distribution Systems (GDS) and also by taking more bookings through its Central Reservations Office (CRO).

As a hotel chain, Thistle Hotels has got a state of the art Website with many features, including online, real-time bookings. Thistle Hotels experienced very significant increases in Internet sales from 2000 to 2001, i.e. nearly 200%. For comparison, GDS bookings increased by just 15% during 2001 in average. In the beginning of 2001, revenue from Internet bookings ran at 92,000 £ per week, up from 27-30,000 £ per week in year 2000. In terms of Internet generated revenues, Thistle Hotels ranks highly among hotel chains in the European market.

Mr. Euan Michell is Thistle Hotels' director of distribution, in charge of all electronic sales and information platforms such as GDS, Internet, Extranets, WAP, PDAs. He has been with Thistle since June 2000.

Thistle Hotels does not offer any discounts for online bookings. The price is the same in all channels. Internet discounts may be a high-risk strategy, but the arguments are strong: It may give incremental business. And there is cost savings to be made of $4 per Internet booking. However, Thistle Hotels is cautious not to alienate the travel agents. Internet booking transactions used to be via Pegasus, but are not taken directly by Thistle: This facilitates data capture, gives Thistle more information, at less cost per transaction.

Thistle does offer E-Deals, for specific hotels, with up to 50% off the rack rate. An example of E-Deals is two nights at the Thistle Charring Cross, for £99 per room night including a full English breakfast. Online booking runs smoothly:

"Book online now - just enter 'STE' into the promotional code box under the 'Special Offers' category."

Content of the online personal brochure:

- General hotel information. Address, description (p. 1).
- Directions: Rail, air, road (p. 2).
- Leisure (p. 3-4).
- Business facilities (p. 5).
- Conference and Banqueting, including detailed floor plans for meeting rooms (pp. 6-21).

Thistle Hotels have partnered with Multimap.com to provide online mapping for its hotels. The driving directions facility can for example provide driving directions from Stanstead Airport to Thistle Charing Cross. It is a trip of 47 minutes and a distance of 35 miles. Some people may prefer alternate means of transport such as train and tube, also briefly described on the site.

The Extranet includes corporate rates for hotel rooms etc. It is not being used for bookings.

8.9.3 The WAP site of Thistle Hotels

Thistle's interest in WAP started early in 1999. The WAP site opened in April 2000. To some extent the WAP site contains the same information as the Website. However, there is no booking function on WAP at the moment, but it is possible to click through from the WAP site to the reservation centre, via the use number function. Thistle's Central Reservations Office (CRO) handles all telephone bookings. This call centre is manned 24/7. Differing time zones is one of the reasons why it is an advantage to centralise bookings in a CRO.

It is possible to search for hotels by region on WAP, as on the Web. There are how-to-get-there by rail, air and road directions for all the hotels and lots of other text based information on WAP, such as check in/out instructions. Also, Thistle have offered a Valentines Special on WAP. The WAP site is mainly for business travellers, but for leisure travellers as well, and with leisure offers to business travellers.

Initially it was possible to lock in on WAP and make availability checks, but this function has been disabled since there was difficulties with big files over WAP, minimising the file size solved the problem. Also speed of connection remains a problem.

Participation in mobile portals was considered, but it is relatively expensive. M advertising is still an embryonic field. It is difficult to justify the high costs of advertising on the main portals of the medium, for example M-viva of Mobile Warehouse and Genie of MMO2.

It was the intention to put Thistle's share prices onto the WAP site. Thistle Hotels will have an in room portal for fixed-line Internet access via ADSL, with a Virtual Private Network thus creating an office in the hotel room for the guests. The retail fixed-line portal will be integrated with WAP. It will benefit WAP when speed improves with GRPS networks. Also, with GPRS handsets clarity of screens will improve. The children of today are familiar with mobile phones: WAP will benefit from the youth market.

Thistle Hotels have appointed a new web agency to further increase Internet bookings. Organic.com thus took over from IXL. The new Web agency is also responsible for the WAP site with one of the criteria for selection being the new Web-developer had to be strong in m-commerce. Thistle received 7 proposals, had 3 presentations, and selected Organic.com.

Although Thistle Hotels is more interested in exploiting the opportunities of the fixed line Internet than WAP, the latter is by no means ignored. Thistle thinks its WAP site is already relatively good, but there are lots of ideas and plans for further development. Personalisation will be a dynamo. Directions about how to get to the hotels will be offered as a Location Based Service, LBS. Also maps are something for the future in this connection. The location aspect is thus considered to be very important, not least on the mobile platform. In future Thistle will offer real-time availability on WAP, like on the web. Booking on WAP will be for those already registered on the Website.

In the future it may even be possible to check-in by WAP. As the guest walks out there may be a sign with a personalised thank you for your visit message. Finally, there could be a service with information about flight delays, if any, and information about the current location or status of the airplane.

8.10 HRS, Cologne, Germany

8.10.1 HRS - The firm

HRS was founded in 1972, as a hotel reservation company. It has a department for meetings, special events and group reservations. HRS target hotels of all categories: individually run hotels, hotel corporations and hotel chains. HRS is a worldwide distribution channel for the hotels, free of charge for the travellers, which are business and private travellers and international companies. HRS negotiate low rates with hotels, which can be updated permanently (Hotel Self Administration tool on the internet available) and earn a commission from the hotels for each booking. The hotels pay a 10% commission, and there is a EUR 250 flat-rate charge p.a. for internal data administration only, when more than 50 overnights have been generated. HRS receives an override commission only, if the turnover amounts to more than EUR 50,000 during the calendar year. Guests settle payment at the hotels. Many companies, more than 9000, use HRS as the preferred intermediary for their hotel reservations. Examples include Allianz, EADS (Airbus Industries), ERGo Group, Bayer AG, Bosch, Electrolux, Nestle and many more. HRS is partner of the leading mobile phone companies and of airlines like Austrian, Germanwings, Swiss, LTU and partner of public interest websites like German railway and many others. HRS has 140 employees. An office in Shanghai, PR China, has been established since 2 years.

8.10.2 HRS' web site

The HRS websites hrs.com and hrs.de were established by the end of 1995. They offered travellers the ability to make online hotel reservations worldwide. website visitors are taken directly to the booking function and there is no redundant information. There is information about distances and instructions about how to get to the hotels and city maps are provided, no credit card details are needed if arrival is before 18.00. HRS counts more than 15 million page views per month and around 80 million hits with the site being available in 24 different languages. Online bookings are increasing; from 1999 to 2001 the increase in the hotel turnover amounted to an average of 68% whilst 2002 finished with an increase of 53,49% (to 2001). In January 2003 there was an increase of bookings of 45,73% (to January 2002).

8.10.3 HRS' WAP site

The first versions of HRS' WAP sites, wap.hrs.de and wap.hrs.com were opened in the spring of 2000. Version 2.0 was opened in the beginning of 2001. There is a special structure for WAP users, but essentially the WAP site comprises the same features and content as the Web site. Only bookable hotels are displayed (on WAP). The search procedure is first by country, then by cities, then arrival date and number of night, hotels with available rooms are shown, then the types of rooms at which stage the booking is complete and a reservation number is provided. The booking is guaranteed until 6 pm only.

Both business and private travellers use the WAP site and neither of these groups is targeted specifically.

The WAP site was developed internally as an i-mode version too. The WAP site should mostly be similar to the Web site. It should contain no special information i.e. no information that is not available on the Web site. Focus was on the essentials and emphasis was on having minimal clicks, the WAP booking procedure was kept short at less than two minutes. The WAP site in its several forms has been tested on multiple phone handsets, including Nokia, Siemens and Motorola.

GPRS is keenly awaited, which will give the site more speed, but for now HRS is happy to only offer WAP as a niche service.

The hotline phone number is on the WAP site too so it is possible to book by telephone (voice). It is too early to do detailed user studies; HRS is interested in making extra incentives for those making online and WAP bookings but ultimately just see WAP as an additional way of making a booking.

In addition to reservations on WAP, other supplementary WAP-services in the future will mostly be standardised types of information such as flight schedules, stock exchange data and items relevant to the daily needs of individuals and businesses. As the installed base of WAP enabled phones increase will users actually using the HRS WAP site also increase? Certainly WAP services have to be kept simple and effective if they are to be used.

8.11 Lastminute.com, London, UK

8.11.1 The firm - Lastminute.com plc

Lastminute.com was founded in April 1998 by Brent Hoberman and Martha Fox. It went public in March 2000 and is quoted at the London stock exchange. The firm has 600-650 employees, following the acquisition of

Degriftour in France. For official numbers, please see the published quarterly and annual accounts.

8.11.2 The web site

From the outset, Lastminute.com has been focussed on the Internet as a sales channel. Lastminute.com is a premier site for last minute services.

Mr. Babak Fouladi was Head of New Platforms and Strategic Planning, it was his responsibility to look into new distribution platforms in order to get the deals to the consumers effectively.

Lastminute.com is one of the largest e-commerce Web sites in Europe. - Lastminute.com generate lots of non-travel transactions: Entertainment tickets, gifts and more. These non-travel services and items tend to be relatively low priced, therefore most of the sales on Lastminute.com are travel related.

8.11.3 The WAP site

The WAP site was established as early as November 1999. Lastminute.com was one of the first services on Genie, the mobile portal of the mobile arm of British Telecom, now MMO2. There were not many WAP phones in use at the time of the launch.

Basically, the same products are offered on WAP as on the Web. Same content but different layout and programming languages: WML and HTML. The target groups and users are the same on the Web and WAP: ABC1 are often WAP users. ABC1 is the main target group but the products are for everybody. Users are both business and leisure travellers. When GPRS arrives it is expected that there will be a wider range of users (possibly with more business users).

The WAP site was developed internally and it was relatively easy to develop. With XML the focus is on architecture. Content can be presented in different formats, according to different style sheets. Once this is in place it is easy to adapt the content and layout to a new platform. Personalisation will be important on WAP: The WAP site becomes **your** WAP site. Voice recognition offers interesting opportunities as does the integration of WAP and voice recognition.

Lastminute.com developed WAP on 2G. It did not wish to wait for 3G before developing WAP services and mobile information services. Lastminute.com wants to offer Location Based Services (LBS), although operators do not provide Cell ID information yet there are smart ways of getting the location of the user. Although Lastminute.com is not happy with WAP as a technology at the moment, there are many things that can be done. Making use of SMS is one possibility: International SMS i.e. SMS used in combination with WAP.

Lastminute.com has a popular WAP site and receives a good number of hits. Lastminute.com is everywhere, on all the mobile networks and mobile portals in the UK, France and Germany.

In the hotel field Lastminute.com has made special deals with Bass Hotels and Starwood. It will become possible to make payments on WAP soon, the phone is a personal device that travellers always have and personalisation will be a key buzzword. When you travel, your phone or PDA will be your companion for travel services.

WAP usage is on the rise. WAP 1.1 was just the beginning, lots of people have to work together to enable and implement WAP services: Handset manufacturers, network operators and content providers such as Lastminute.com. Star trek will become reality; users can ask wearable computers questions and get answers. Soon there will be no need to scroll on WAP, just talk to navigate. Phones will change by 2005 or 2006 - PDAs and phones will merge as one and will become wearable. Bluetooth will arrive and mobile payment methods and digital signatures will become standard. It is important to be excited about future possibilities but it is also important to do something now.

Customers are valuable, if you annoy them they will not come back and they should sign up for what they want. The phone is a personal device and only relevant information should be sent to mobile phones via SMS. It may be fruitful to combine voice, SMS and WAP.

8.12 LeisureHunt, UK

8.12.1 The firm

LeisureHunt was founded in 1996 as a showcase for Whereonearth.com, which offers digital maps of the world, with coordinates for the different Points Of Interest (POIs), for example for local governments and utilities. All the POIs are geo-coded, including B&Bs (Bed and Breakfasts), guesthouses and hotels.

LeisureHunt.Com had been operating as an independent company since October 1999, but 15% was still owned by the owner and founder of Whereonearth.com. - On 1st November 2002 it was announced that World Travel Holdings plc (WTH) had acquired LeisureHunt.com Ltd. Mr. Tony Prior, the CEO of LeisureHunt was at the same time appointed to the board of WTH as group marketing director.

The main core of the staff is based in Ipswich, Suffolk with another office located in London. LeisureHunt has 21 employees, including 7 on the reservations side for taking phone calls as well as online bookings. Less than

5% of the bookings are telephone bookings and the rest are Internet bookings. LeisureHunt does trade marketing only, i.e. no marketing toward consumers.

LeisureHunt's focus for the future is to become a complete accommodation service provider to other major Internet service providers and Internet information sites. The company's approach is to connect as large a consumer audience as possible with its hotel database via Web, WAP and iDTV distribution partnerships. Mr. Nigel Muir was Director of Business Development, in charge of utilizing any potential technology including Web, WAP and Interactive Digital TV.

8.12.2 The web site

The website was established in 1997, and was re-introduced in March 2001 on a new platform in connection with a change of hosting arrangements. Now the content is XML enabled and there is a better interface.

LeisureHunt's hotel database includes 100,000 properties worldwide, of which 31,000 are in the UK and 69,000 overseas. LeisureHunt uses the GDS Amadeus for some of its availability data and content for chain hotels. On top of these come non-chain hotels. All the 69,000 non-UK hotels are online real-time bookable. There are 5000 UK online real-time hotels. The rest of the UK hotels, 26,000, are online bookable on a request basis only. Commission is shared, with 2 Euros per booking to Amadeus with a 10% commission from the hotels.

LeisureHunt's technology enables a prospective customer to search for a particular property within a specific location, in a certain price range and with specific facilities.   Its search facility, based on the Whereonearth.com database, recognises and locates 2.7 million place names worldwide that are geo-coded on a global digital map. Because all 100,000 properties are also geo-coded the search engine can identify all of those within a selected distance from the start point. The search facility will continue rolling out from the destination point until it finds a meaningful number of hotels to offer the user. Apart from LeisureHunt.com, the customers of Whereonearth.com also include Lastminute.com, amongst others.

Of the Internet bookings 60% are real-time bookings and 40% are on a request basis. 80% of the bookings are made by UK customers and 20% ae from overseas, including USA, Japan and continental Europe, Australia and New Zealand. All the revenue from LeisureHunt comes from hotel bookings.

LeisureHunt currently has a 2% conversion rate (from website visitors to online bookers). This rate is considered ok, but LeisureHunt is aiming for 4%. LeisureHunt offers property management systems (PMSs) for small properties, the PMS is called speed book and at the moment it is used by 500-600 UK hotels.

LeisureHunt counted 1.1 million page impressions per month and 80-90,000 unique visitors each month in 2001. This was a 200% increase over the year 2000. Floating on the stock exchange was considered, but postponed.

LeisureHunt will now focus on the UK market; outside of the UK its offer is not compelling. LeisureHunt will collaborate with companies outside of the UK, though, and sees big opportunities in Continental Europe.

To successfully integrate the service into leading brands' Internet sites, LeisureHunt develops and operates private label versions of the service as well as co-branded sites. LeisureHunt is represented on major web sites such as Yahoo!, Freeserve, ICircle, Lycos, Ask Jeeves, Emap Online (a2btravel & escaperoutes), Virgin Net, Amadeus.net, and Supanet.

12.3 The WAP site

In the beginning of 2000, WAP became a must have for hotel intermediaries. LeisureHunt's WAP site went live in March 2000. Its first partner was Genie, then Mviva and Virgin Mobile. LeisureHunt believes its WAP site needs more development. Previously the WAP site returned the phone number of the hotel, now the phone number of the call centre is listed on WAP, which is not an ideal situation. LeisureHunt would like to establish a new revenue model with the key questions being: Where is the revenue stream and who should pay whom?

LeisureHunt currently devotes only limited resources to WAP, it does not invest too much in WAP at the moment. LeisureHunt is looking forward to location sensitive services, LBS, which would allow for example the provision of directions i.e. how to get to the hotel from the current or another given location. In the future poeple will have multiple devices and LeisureHunt is thus talking to PDA-oriented firms like Avantgo, Palm and Vindago.

There is currently only a small ratio between page impressions (PI's) on the web and WAP, i.e. relatively few PI's on WAP compared to the Web. In 2001 500 unique visitors and 1700 PI's per week were counted. Lots of WAP site visitors are leaving shortly after they have entered the WAP site.

Whereonearth.com developed the WAP site internally; it was tested on the Nokia 7110 and on an Ericsson handset. LeisureHunt's WAP site could not run on Freeserve's WAP emulator, which delayed the launch. LeisureHunt gave up on the Nokia 7110 on 2G and is now looking forward to GPRS.

As to their perspectives on WAP, they see location-based services (LBS) as an interesting opportunity. At the moment, in order to pinpoint the location, the user has to type in the junction of two streets, which is obviously cumbersome whereas in the future the current location of the mobile phone will be known provided the mobile network operators release this information.

At the moment hotels with rooms that are not available are shown on the LeisureHunt web site, which leads to some frustration among consumers. In future LeisureHunt will show only hotels with available rooms on its website. The same must be the case with WAP. A WAP phone is not the right device for prolonged searches and is only for quick, accurate information.

Other useful services on WAP may be: Find me a taxi and flight information. SMS can also help: Criteria can be set up on the fixed-line website and information can then be sent out by SMS according to a profile. For business travellers it may be useful to know the time of the next flight back to headquarters. Railtrack went into receivership, but train related information is also appropriate for WAP. Perhaps 30% of those who have got a WAP-enabled phone have tried to use the function. There are more people who subscribe to SMS alerts than there are WAP-users.

The WAP site of LeisureHunt is no longer live.

8.13 Check-Inn.com, Germany

8.13.1 The firm

Check-Inn.com AG was founded in June 2000 as a company, but the Check-Inn.com website has been live since January 2000. The company has about 10 employees. The business model is the following: Free listing and presentation on the web and WAP with no fixed annual fee, just a booking fee for each Web booking. Check-Inn.com also earns revenue from online advertising. Mr. Carsten Strobel is CEO of Check-Inn.com, mostly focusing on technology; he and a number of other key persons handling technology and marketing are based in Aalen, whereas the financial side of the firm is handled from Hainburg, by TWS Treumandat. Check-Inn.com offers a business-to-business service to hotels, namely joint purchasing, like the firm DMM International Trading Co., Inc., which is a majority shareholder of Check-Inn.com AG.

Since March 2001 Check-Inn.com has offered a service to registered members called Hotelhelp, which means that Check-Inn.com guarantees that it can find an available bed/room in or near a given city, even during exhibitions and conferences, when hotels are particularly hard to find. This service is offered at 99 Euros per month. In the beginning of 2002 it was announced that Check-Inn.com made a profit during 2001, its first full year of operation. For the year 2002 an increase in revenue above the (online hotel booking) industry average was expected.

8.13.2   The web site

Check-Inn.com's online database contains more than 600,000 hotels, 30,000 hotels are listed with availability. Check-Inn.com liaises with many suppliers for availability information. Thus it gets this information directly from hotels, from Worldres.com and from a front-desk software application. Check-Inn.com is not charging the hotels any fixed annual fee, unlike some other hotel intermediaries. Everything is based on transaction fees, i.e. there is a commission fee per transaction, there is also a fee for priority listings of hotels on the web and WAP.

Check-Inn.com wants to create the largest and best source of hotel content on the Web; it is a sort of content management system for the hotel industry. The

website lets customers find hotels, and book them. It contains the classical search features. And there is a route planner on the site.

Check-Inn.com has an affiliate programme, like amazon.com, which in this case means that web site owners, who are members of the affiliate programme, can earn a commission for Check-Inn.com hotel bookings made from their web sites.

Hotels increase their capacity utilisation because of their participation in Check-Inn.com. There were 70,000 bookings in December 2000, on the Web, with 1.4 room nights per bookings. For each booking made through a middleman such as Check-Inn.com, there are 10 bookings directly to the hotel.

The website was re-launched in June 2002.

8.13.3 The WAP site

Check-Inn.com has had a technically functioning WAP site since June-July 1999 (developed on an emulator). In March 2000 the WAP site was officially launched. The goal was to double the WAP traffic every two months. In December 2000 Check-Inn.com counted over 1 million page impressions on WAP. Check-Inn.com's WAP-service is thus quite popular. It is on the portals of both of the main mobile network operators in Germany, D1 (T-Motion) and D2 (Vodafone). In the mobile area, Check-Inn.com co-operate with Ericsson and others such as the major mobile network operators in Germany and beyond. At the moment it is hard to use the WAP system, partly because of increasing traffic. The WAP site comes in four different languages: German, English, French and Spanish.

The reason for establishing the WAP site was to offer the best possible service for the user. However, it is hard to see why people should want to do everything on WAP. It is better just to let people call the hotels directly, and let them make their room booking requests by voice on their mobile phone.

In terms of generating revenue, there is not much to do. Check-Inn.com has discussed revenue sharing with the mobile network operators. It is, however, difficult to reach an agreement with mobile network operators in Europe about sharing airtime revenue. For Check-Inn.com it's nice to have a WAP site and they are happy to have it, but the WAP site is not generating any revenue for Check-Inn.com yet. Nobody earns money on WAP. Many are - or were - pushing it. But it is unlikely there will be any revenue until 2003 or 2004. WAP needs more time, at the moment nobody can be charged any money for the WAP-service usage, except for the airtime, the revenue of which goes to the mobile network operators. Currently WAP costs Check-Inn.com money, so really the WAP site should be taken off-line however Check-Inn.com stays in the WAP business in order to try and be at the leading edge.

Check-Inn.com developed a Location Based Service for D2 (Vodafone in Germany), and became the first hotel site in Germany to offer location-based services (LBS) on WAP. An increasing number of its hotels are geo-coded, in Germany alone 13,000 by the end of 2000. Check-Inn.com wants to get away from fixation on cities. Also fairgrounds are geo-coded, as points of interest (POIs) in the system. Surprisingly, neither of the key words ICC or Messe is recognised as Points Of Interest. It is possible though, for example, to get a list of hotels close to Brandenburger Tor or Berliner Zoo. In the Berlin LBS solution there were 370 hotels and over 2100 other POIs including theatres and restaurants from the outset in February 2001, i.e. just before the annual ITB exhibition, which takes place each year in the beginning of March at ICC in Berlin. By September 2002 there are a total of 764 accommodations in Berlin listed in the Check-Inn or hotelkatalog.de WAP sites. Check-Inn.com is realising geo-data on a regional basis. Apart from Berlin the geo-coding of POIs has also been undertaken for Friedrichshafen, another trade fair city. The Berlin solution was originally at wap.btm.de, then atwap2berlin.de, hosted by daland.com. Daland is one of about 50 German members of Ericsson's Mobile Applications Initiative.

It would be possible to establish a booking function on WAP. But booking on WAP is too complicated. Today there are the phone number, fax number and e-mail address, if available of each hotel on the website as well as on the WAP site of Check-Inn.com. On the WAP site there is a link to call the hotel. Calls cannot be protocoled though, only bookings over the website can be counted.

Hotelcatalogue.com of Daland in Berlin is affiliated with Check-Inn.com in Aalen, Germany. Mr. Joachim Kastner of Daland is a shareholder of Check-Inn.com. The hotel database of Daland's site called Hotelcatalogue.com and that of Check-Inn.com is the same. Daland focuses on WAP and handles the technical side of the WAP site of Check-Inn.com (wap.Check-Inn.com/), whereas Check-Inn.com focus on the fixed-line web site. Daland had the expertise in WAP and developed the WAP site in-house; a new service was created for WAP since the usage process on WAP is different from the web.

Check-Inn.com targets all types of travellers, on the web and WAP alike, and on neither of the platforms is registration of users required. To get WAP bookings going, there must be a discount offered. WAP bookings must be possible with a few clicks only and the user should be easily identified. A secure payment system on WAP is required and payment is take from the users phone bill.

WAP usage is small compared to Web usage. Normal level of Web:WAP usage is 20:1, i.e. 20 times as many page views on the web as on WAP. Check-Inn.com is present in many, almost all, WAP portals in Europe. D2 launched the WAP service via TV. D2 consider Check-Inn.com on WAP to be a highly desired service. Check-Inn.com is, or at least was, the most used service on D2, according to D2 itself.

With LBS (Location Based Services) there will be no need to tell the system where you are, just push a button on the WAP phone. The location will be known from data provided by the mobile network operator. Plans for further hotel related WAP developments include the provision of driving directions, for this purpose a rather exact location is needed, at the moment the accuracy is within the range 200-300 meters yet it is expected in the future it will be within 10-20 meters.

For those without a WAP-enabled phone or for those who do not (wish to) use WAP, SMS services can be offered. Check-Inn.com already provides driving directions on WAP, i.e. how-to-get-there information. In the future there may be a sort of one stop shop, with train, hotel, change of flight, car, and hotel all offered in one place.

Broader bandwidth for the mobile networks is required, Check-Inn.com has hotel photos at hand and it would be nice to display them on mobile devices. There may be personalised offers, where people have to register, and they will then only get the service or type of information they are interested in.

At the moment only a few services on WAP make sense. For example information about traffic jams is useful on WAP. Also, finding a hotel, now (as opposed to next week) and What's On Tonight. The phone companies should offer those services that make WAP attractive, for example Location Based Services - services relating to that are interesting, but service is slow today.

Check-Inn.com thinks it is a mystery that SMS became such a success, almost overnight, i.e. during year 2000, given the cumbersome keying in of letters. It is getting better with T9, though, i.e. predictive texting. Young people are crazy about SMS, they are not hotel bookers, just chatters though. In principle it is the same thing, keying in information on a device which is not best suited for it, it is anticipated the SMS-generation will be users of services such as what's on tonight etc, not hotel booking.

By the middle of 2001 it was thought that 75% of new mobile phones sold were WAP-enabled. Many people try WAP, but are then disappointed. Less than 5% of the owners of WAP-enabled phones are regular users, using WAP more than once a week.

Check-Inn.com has five different WAP-phones in the company, which are used by Mr. Strobel for testing purposes. They include models from Siemens, Nokia, and Ericsson. Check-Inn.com checks which phones visit its WAP site, for example which language versions. German language is dominating among the Check-Inn.com WAP site visitors. Users get the appropriate language version of the WAP site. Check-Inn.com is attempting to be accessible by all WAP phones and PDAs, however, very few PDAs are in use at the moment.

8.14 Venere, Rome, Italy

8.14.1 The firm

Venere Net SpA was founded in 1995 as Italy Hotel Reservation by four people, who knew each other from university, where they studied and undertook research in physics.

Currently there are almost 40 people employed by or affiliated with Venere. Originally the firm was a S.r.l., now it is a SpA.. The firm has not yet made an IPO, floating on the stock exchange, but the formalities are in place. It would be possible go public by 2003 or 2004. Venere Net SpA is funded by venture capital, from Tiscali, amongst others.

Other from Italy, Venere is also present within subsidiaries in France and the UK. In the future there will also be a subsidiary in Spain and one in Germany.

8.14.2   The web site

The Venere web site, venere.it and venere.com, was launched in early 1995. It was one of the first web sites which went live in Italy. When the website was opened and the firm founded, Internet penetration was low. Internet access was mainly by Americans, so it made sense to offer a service of interest to Americans, namely hotel booking. It is possible to search all Italian hotels on the site, from 1 to 5 star rated and book some of them online. There are no intermediaries: There is a direct connection between guest and hotel.

Venere.com lists 2000 hotels in Italy and is the number one hotel site there. Currently Venere counts about 20,000 unique visitors per day. And almost 10 times as many page-views. Back in 1999 Internet revenue was 7 million Euros. In 2000 it more than doubled to 15 million Euro and during the first six months of 2001 gross bookings amounted to 13 million Euros. Shortly after Sept. 11 it was announced that Venere.com expected to reach 25 million in gross bookings for the year 2001, although the negative effects may have been greater than originally expected since a large proportion of the web-site visitors are Americans. No other financial announcements have been made available on the website of Venere.

The hotel owners themselves maintain the web pages, Venere just manages the system and contact is directly between the user and the hotel. This is the main difference between this site and others, this was a new approach, Venere is not like a travel agent or a tour operator as the user pays at the hotel, on arrival or by the end of the stay. This is convenient for both the user and the hotel owner. The hoteliers pay for use of the system, there is no commission, however Venere is changing from a flat annual fee to a 12% commission basis.

Venere is for registered users, of which there were 200,000 by 2001, up from 127,000 in 2000. Website visitor country split is 45% USA, 14% UK, 11% Italy, 17% Rest of Europe and 13% Rest of World. There are 6300 online bookable properties.

8.14.3 The WAP site

The WAP site was opened in May 2000. Before that a pilot test was conducted, for checking the possibilities of the new service. A partnership was formed with phone.com but Venere was also in discussion with others. An explosion in WAP usage was expected, but it did not happen. Currently WAP is a clumsy technology, perhaps later this will improve.

The WAP site is for business travellers and for independent travellers alike, as with the Web site. The WAP site was developed in-house, it was simple, dynamic files were simply created in WML. The scale (or form) factor (small screen size) had to be taken into account however as the WAP site had to be simple to use, quick, and stable: It is usable by different browsers and different screens although there is a problem with compatibility, as transforming content so that it displayed correctly on different mobile phones and browsers proved difficult.

In the future, PDAs may become a more useful platform than mobile phones, with bigger screens. There may also be a call function on the WAP site in the future. Venere is not working on the WAP site it at the moment, there are very few users with the ratio of WAP:Web users being less than 1:1000.

For hotels it is important to have a WAP service though, which lets users find hotels and check availability. Hotel booking on WAP is more useful than other WAP services, particularly because booking hotels is connected with mobility and travel, particularly with last minute hotel bookings.

Venere will increase its WAP presence in due course (this was always the intention). Venere had its WAP service live before the phones were on the market in fact, developing the WAP site on an emulator.

There are many WAP phones on the market (in use), but actual WAP usage is low. It is difficult to configure WAP phones and it is too difficult to use them in may ways therefore WAP is not being used heavily. The mobile telecom operators oversold WAP as they were keen to earn money and they were trying to create closed gardens, trying to keep customers. This was a mistake, the WAP users - like Web-users - do not want to be confined.

The WAP site of Venere is no longer live.

8.15 CitiWiz.com, Paris, France

8.15.1   The firm and the web site

CitiWiz, the handy city guide, was founded in the beginning of 2000. It is based in Paris and has 6 employees. Furthermore the firm has over a hundred freelance journalists working for them, typically located in the cities they are covering. The journalists may have written for Lonely Planet etc.   It is envisaged that there will be no more than 10 employees eventually.   Mr. Jan

C. Berger is CEO of CitiWiz, while Mr. Ollie Gilly is VP of Business Development. The parent company of CitiWiz is a firm called jUMTStart Ltd. The same people are employees of both CitiWiz and its parent company, except the journalists.

CitiWiz currently has partnerships with a dozen wireless network operators and mobile portals. CitiWiz is a premier mobile city guide designed for mobile Internet users who can access relevant pieces of accurate information swiftly. CitiWiz is focused entirely on the mobile medium. It does have a web site, but this is just for providing information about the firm and its WAP services. The website also features a WAP emulator which lets Web-users see some of the content of the city guides that are made available on WAP and i-mode.

8.15.2   The WAP site

The CitiWiz handy city guides enable WAP users to access a broad range of city-specific information in over 100 cities around the world. This makes CitiWiz one of the world's premier mobile city guides. By the beginning of 2001 there were about 75 cities in the CitiWiz system. By October 2001 this number had increased to 106 city guides on the WAP site, and that remained the number of city guides in the system by October 2002, so of course the addition of further cities has been put on hold for a while. Citiwiz is concentrating on Europe.

Each of the CitiWiz handy guides typically contains the following categories of free information services: Hotels, Restaurants, Bars/ Clubs, Car Rental, Shopping, Airlines, Useful numbers and Sightseeing. Each category consists of several sub-categories that give users access to relevant pieces of information.

Citiwiz counted 500,000 hits per month in 2001 on its WAP site. With about 12 hits per session that corresponded to 42,000 user sessions per month. The split based on hits is probably 3% Web 97% WAP, i.e. a ratio of 1:33 in favour of WAP, since as mentioned the Web site is just there to provide information about the WAP site. Citiwiz wants to concentrate on telephones. Whereas Citiwiz noted a great increase in WAP usage from 2000 to 2001, up to 20% per month, the increase was quite small from 2001 to 2002. In January 2002 Citiwiz counted 22,000 user sessions, i.e. an increase of 10% over January 2001. Citiwiz counted 480,000 hits (40,000 sessions) in December 2002, and looks set to reach 424,000 hits (or 35,000 user sessions) in January 2003, according to CEO Jan C. Berger (20th January, 2003).

The Citiwiz WAP guides are for all WAP phone users, from business travellers to backpackers. Loading time is quick: 1.8 seconds. Connection time is fast and navigation is easy. Citiwiz has a zero error rate on downloads. This is achieved through extensive user testing; the site can even display Chinese and other characters. Content is translated to multiple languages including Greek, with the Olympics in Athens 2004 in mind.

Citiwiz generates most of its revenue from mobile network operators. Citiwiz may establish m-commerce, i.e. m-booking capabilities on its WAP site with the business model in Japan being particularly interesting.

The content is written in XML which means that content can be scaled, and easily adapted for WAP (WML) and cHTLM (i-mode) alike.

One of Citiwiz' competitors was CitiKey of Stockholm/London, which is now out of business, although it received considerable funding : CitiKey laid off all its 90 employees in November 2000, as the company entered liquidation. It tried to put everything into its guides. e-street.net acquired part of Citikey's assets in February 2001, some other players in the field of city guides for mobile phones and other wireless devices are still in business, for example Wcities.com, max.de/cityguide and city-guides.ch. Wcities.com lists 300 cities, including 129 in Europe and 90 in North America. Max.de/cityguide covers just 33 cities, of which 10 are in Germany and 17 in the rest of Europe.

Max.de/cityguide counted as many as 700,000 users sessions in January 2002, possibly to some extent driven by a popular fixed-line web site, which runs in parallel with the WAP-version. city-guides.ch concentrates on Switzerland only and is described elsewhere.

CitiWiz is aimed at 25-35 year olds and the WAP site was developed in-house. It has a carefully thought out structure, with a low click depth. There is no search engine, the members of the CitiWiz team are proud of their WAP site but it is still early days. They still wish to expand content and they have received positive feedback from users. They aim for a zero error rate.

8.16 Wonderful Copenhagen - WoCo

8.16.1   WoCo - the organisation

Wonderful Copenhagen is the official tourism organisation of the Greater Copenhagen area. Wonderful Copenhagen® has it's roots back in 1887, when the first tourism organisation of Copenhagen was established. Today Wonderful Copenhagen® is the official convention and visitors bureau of Greater Copenhagen, dealing with all aspects of Copenhagen as a tourist and travel destination: Marketing, promotion, product development, strategic planning, information, PR, brochures, statistics, tourist information office, analysis etc. Wonderful Copenhagen has about 60 permanent employees. In 2000, the budget was DKK 90 million (12 million Euro), of which 39% was publicly sponsored.

8.16.2 WoCo's web site

WoCo opened the first version of its website in 1996. In January 2001 it was redesigned. Visitor numbers have been increasing steadily, but there are peaks in connection with major events, such as the European Song Contest which took place in Copenhagen in May 2001. 25% of the website visitors are Danes and the rest are mainly from the USA, Germany, UK, Norway and Sweden. Website stats for Visitcopenhagen.dk and several other Danish tourism and hotel-related sites are published on a monthly basis, i.e. hits, page views, user sessions.

For Visitcopenhagen.dk, WoCo's web site, a user session was defined as only 15 minutes until the end of 2001, but from then on 30 minutes, as with the two other sites. To some extent this explains the lower number of user sessions per month during the first half of 2002, compared to 2001. However WOCO believes that the main explanation is a change in their website which had the side-effect that it became less likely to be found by search engines, but after doing some work on placement in search engines and SEO traffic is picking up.

8.16.3 WoCo's WAP site and SMS info service

WoCo opened its WAP site on 1 July, 2000, with attractions as the main content. The menu of WoCo's WAP site contains the following options:

- Attractions (156 in total, in 5 categories)

- Copenhagen Card

- Tourist Office

- Hotel Booking

Attractions: Search for specific attractions or get a list of all attractions within a specific category. Copenhagen Card: Information about the Copenhagen Card, which gives you free access to more than 60 of the most popular museums and attractions in Copenhagen. Tourist Office: Contact details for Wonderful Copenhagen Tourist Information in Copenhagen. Hotel Booking: Information on who to contact if you need accommodation in Copenhagen.

The attractions section is the main content. Attractions can be searched by name or by category, one of which is Nature, Zoo's etc. There are 13 entries in this category, one of which is Denmark's Aquarium. One click and address, phone number and website address is shown.

Entry is free with a Copenhagen Card, and so is public transport in the metropolitan area of Greater Copenhagen. The transport section explains different ways of getting there, by combinations of S-train or regional train and bus. If you click on Hotel Booking you simply get the following information:

Call +45 70 22 24 42

Or visit the Wonderful Copenhagen Tourist Information

At the call centre in the WoCo Tourist Information office, they keep an updated database of available hotel rooms; WoCo earns a commission on hotel bookings. WoCo counts 600-700 user sessions per month, i.e. visits to the main menu page, on the Danish version of its WAP site and about 400 per month on the English version, i.e. about 1050 WAP

The information on the WAP site is the same as on the website (Copenhagen Card, attractions and sights). The call centre updates the web site and at the same time the information is put on the WAP site. The underlying data which is used both for website and the WAP site is kept in an Access database. During the WAP site design process the need for user friendliness and usability was kept in mind.

Apart from its WAP-service, WoCo offered visitors the possibility of subscribing to an SMS info service during the course of their stay. Visitors could choose to get SMS's about one or more of the following categories, by checking the appropriate boxes at a certain registration page on the WoCo fixed-line web site: Music, sports, activities & entertainment, exhibitions, gastronomy, shopping. The SMS service was free of charge.

The above service has now been shut down. For 2003 WOCO will try something new, Copenhagen has a free city map, where a number of attractions are marked. Each attraction has a number - from spring 2003 visitors can SMS CPH xxx (number of attraction on map) and send it to 1266 - they then receive an SMS message with information about opening hours and events at the attraction - the service costs DKK 2 per SMS.

It may be a problem that foreigners cannot necessarily access WAP sites in the country they are visiting. Roaming is a problem area and there may be a need to reconfigure the WAP phone when moving from one country to another. WoCo cannot advise all foreign WAP users about how they should configure their WAP phones, the traveller is required to contact his or her mobile network operator regarding this.

WoCo is rather satisfied with the functionality and content of their WAP site. The content is good and simple, sensible, short and precise information. No jargon.

Visitors have normally made their hotel booking from their home base, via the Web or their travel agency. On WAP it is therefore more relevant to provide information about what's happening (what's on) when the visitors have arrived and maybe information in relation to tasks such as booking a taxi.

From 1 January 2003, WOCO has shut down the access base on which the service was based and therefore the WAP service has also been shut down.

8.17 Danish Tourist Board

8.17.1 Danish Tourist Board - the organisation

During 2001 the Danish Tourist Board (DTB) had 128 full time employees, in average. The spending budget was DKK 197 million (Euro 26.5 million), like the year before. 75% of the annual budget is covered by governmental grants, 25% by industry as contribution to marketing activities.

8.17.2 The website of DTB

DBT opened the first version of its website in the beginning of 1997, but the underlying product database, originally called Dandata, has existed since 1985. Cap Gemini was selected as the main technical contractor in November 2000 following an EU public tender to revamp the DTB consumer web site. The site was re-launched on 22 April 2002. The website of the Danish Tourist Board comes in 12 languages and contains 1 million pages. In the 2001 budget,

DKK 22 million (Euro 3 million) was set aside for Internet portal activities of which DKK 17 million was set aside for technical development of fixed-line Internet portal visitdenmark.com. The re-launch was postponed from late 2001 to April 2002 and therefore part of the planned payments was postponed accordingly. - There are about 2700 locations in the system, including all 974 Danish hotels with more than five rooms.

8.17.3 The WAP site of Danish Tourist Board

In January 2001 DTB launched a WAP site in three languages. The site was developed in cooperation with TDC Mobile (TeleDanmark) and hosted by Payfish. As Payfish went out of business in September 2001 the DTB WAP site was discontinued.

Out of a total of 42 types of information (products), 8 or 10 were chosen for the WAP site :

- Hotels

- Camping Sites

- Youth Hostels

- Attractions, museums

- Activities

- Events

- Music

- Ferries

- Tourist Bureaux

All the 1000 hotels which are on the DTB website were also on the WAP site, they could be searched for by name, town and number of stars. The primary target group for the WAP site was originally business travellers, DTB did not promote WAP apart from issuing a single press release. At one point in time WAP was heralded as the next big thing, but it swiftly went downhill, DTB intended to implement a welcome message by SMS for visitors arriving from abroad in cooperation with TDC Mobile. This should be implemented with a link to the WAP site of DTB. This idea was never implemented however.

It was not very expensive to develop the WAP site, it was developed externally by Payfish under the supervision of DTB. TDC Mobile recommended Payfish although there was no real specification of requirements. Core tourism services (products) were to be included and then services related to mobility as well as the tourist bureaux.

XML will be implemented in the future; this provides the possibility of channelling information to different devices. There was nothing complicated in the the development of the site, it was just a question of making a wish list at which point the prototype was built very quickly.

WAP was not related to the fixed-line portal project and its marketing, the WAP site was more a way of showing that some development was going on until the major fixed-line portal project was finished. The WAP site was considered to be acceptable, DTB did not have any development plans for WAP except the implementation of XML that would facilitate easier updating of the WAP site (if it had been continued). Apart from this it was the intention of DTB to wait for the next generation of mobile services.

WAP is not considered by DTB to be of much use yet. It is not always possible to access WAP sites abroad and WAP is too slow. The screens are too small although this will improve when PDA's become mobile i.e. equipped with GSM-modules.

8.18 Bedhunter.com - of ehotel AG, Berlin

8.18.1 ehotel AG - the firm

In February 2000, Matthias Kose (an IT and m-commerce expert) and Matthias Garke (a hotel industry expert) set up the hotel reservation service bedhunter.com GmbH. By the end of September 2000, bedhunter.com GmbH was merged with the ehotel service of i:FAO into a company called ehotel AG. This company offers its customers at www.ehotel.de and www.bedhunter.com free reservation services for hotel rooms at attractive rates. Hotel bookings are taken via a call centre, online via the aforementioned website addresses (with identical content), and the WAP site wap.bedhunter.com. ehotel AG has 25 employees at its main base in Berlin and another 10 at its branch office in Frankfurt, plus freelancers. ehotel AG has more than 60,000 hotels in its database. ehotel AG gets 10% commission of the booking value. ehotel AG also earns income from customizing and selling its online hotel booking technology to other companies, including Lufthansa, Deutsche BA and Sixt.

During the first 9 months of 2000, i.e. before the spin off, revenue of the ehotel division of i:FAO was Euro 5.1 million (up 30% from the year before), net result was minus Euro 0.5 million. For the full year 2000 (12 months) revenue for the ehotel division could then have been about Euro 7 million (plus some revenue from the Bedhunter GmbH business). In the annual accounts of i:FAO year 2000 in was stated that "positive results are not yet expected for the ehotel AG in the next two years because of high expenditures related to the opening up of new markets." Originally, 49% of the shares in ehotel AG the were held by i:FAO and 40% by the former shareholders of the bedhunter.com GmbH, and the reminder by industrial investors. During 2001, i:FAO wrote off completely its investment in ehotel AG. In the 2001 accounts of i:FAO, the 49% share in ehotel AG affected the bottom line of i:FAO negatively by Euro 0.88 million. In other words during 2001 ehotel AG was still in the red - as expected - by about Euro 1.8 million. During 2002 ehotel AG increased it share capital without the participation of i:FAO. Therefore, by the end of Q1.2002, i:FAO share in ehotel AG had been reduced to 42%, and by the end of Q2.2002 to 40%. During Q3.2002 (28 August) i:FAO sold all its shares in ehotel AG. I:FAO aims to be a leading developer of eProcurement software for business travel.

8.18.2   ehotel AG on the web

In February 2001 ehotel AG made a deal with Trust International, which gave ehotel (and bedhunter) access to rates and availability data for 40,000 hotels, worldwide, via Trust but from Amadeus. In addition to this rather general hotel data, ehotel AG also has negotiated prices. In September 2002, this deal was replaced by a two-year contract with Dallas-based Pegasus Solutions, which gives ehotel AG access to rates, availability data, photos and detailed property information for about 47,000 hotels, of which more that 1700 in Germany alone. So, now the call centre, the (two) web site(s) and the WAP site of ehotel AG plus the earlier mentioned partners of ehotel use Pegasus' Internet booking engine and database to make room reservations. ehotel AG hasan XML interface to Pegasus' Electronic Distribution Switch which in turn gives access to the central reservations systems (CRS) of more than 200 hotel chains. Thanks to Pegasus, these same hotel chains are also bookable via Global Distribution Systems (GDS) such as Amadeus, but since ehotel gets the hotel data from Pegasus, the GDS' are not involved in the case of ehotel.

8.18.3 Bedhunter.com - ehotel AG's WAP-service and SMS-service

The founders started taking an interest in WAP in late 1999. The site wap.bedhunter.com was actually opened in May 2000. About 15% of all hotel bookings in German hotels are made on the day of arrival. These late bookings are potential WAP (or call centre) bookings (for those who have a WAP enabled mobile phone and use of the WAP function), contrary to hotel bookings made well in advance for which the fixed line Internet service is more appropriate.

After getting Internet access, it takes 20 steps and three minutes to make a hotel booking on the bedhunter.com WAP site. If it took any longer than this customers might prefer to make the booking by voice at the call centre. ehotel hopes that the WAP booking time can be reduced by 50% with GPRS. Steps are: 1) Language selection, 2) select hotel booking, 3) country, 4) select a city, i.e. one of about twenty major cities in the case of Germany 5) select part of city, 6) hotel category, 7) single or double room 8) Dates, 9) Nights, 10) Name and telephone number. Eventually the booking is confirmed by SMS, in the future auto detection of user ID may be implemented.

In addition to WAP bookings, ehotel AG also offers SMS bookings, in co-operation with iobox. By writing the letters "bh" (short for bedhunter) followed by a "city name", and sending it to a certain phone number for one of the mobile network operators in Germany, it is possible in return to get a business hotel recommendation for "tonight". In return a SMS is sent to the guest with the name and address of the hotel, and again the rate.

The WAP site has largely the same functions as the Web version. The WAP site is mainly for the business traveller, but also for leisure travellers. The target group fundamentally is all WAP users, the WAP site was developed in-house and initially there was no documentation about WML programming. The gateways of the operators were not stable and the functions differed from operator to operator. Dr. Kose is not content with the current WAP site as a payment solution is needed, although Bedhunter's wap-site got five stars in the Focus magazine. ehotel AG believe they have a 1:10 ratio of wap : web usage (visits). In one big WAP portal bedhunter was the exclusive hotel supplier, which may explain the exceptionally fine WAP:Web usage ratio. Compared to call centre and Web-bookings, WAP bookings generate very little business, but WAP-bookings are actually received via the bedhunter.com service daily and the technology functions well. Driving directions (on the web and WAP) will be added in due course. ehotel AG have started cooperating with yellowmap about this for hotels located in big cities.

Hotels are ideal for m-commerce. The advantage of booking on WAP, instead of by voice, is that the complete booking can be carried out in one environment. However the booking dialogue must be short, ideally 1 minute. This is faster than any call centre; car rental is also considered a good WAP application as well as flight booking. WAP via GPRS will improve everything with the next step being UMTS. M-commerce will arrive, but nobody knows when. Content providers must create added value for the customer - via m-commerce. Then WAP et al. will truly become a success.

8.19 HOTELdirect.de

8.19.1 The firm

HOTELdirect was founded in 1989 and is based in Bremen, Germany. However, the firm became a GmbH (joint-stock company) in March 2001. The firm specialises in hotel information and reservation systems. Hotel direct currently has six employees; in the future there will be 10 people or more.

HOTELdirect offers actual hotel information by four different media (communication systems):

1.   Web

2.   WAP

3.   Multi-media terminals: Tele Kiosks - for example at railway stations.

4.   Voice recognition enabled computer

This is how the updating of availability data - common for all platforms - takes place:

A voice computing system calls the connected hotels periodically and asks for the vacancies of the day and the two following days as well as for special rates. In case of changes, e.g. unforeseen vacancies, the hotelier can call the system and, by giving his personal identification number (PIN), access the information at any time. The information received from the hotels is stored in a database and is retrievable via different communication systems.

Revenue in 2001 was 400.000 Euros, generated by 200.000 contacts at 2 Euros per contact. HOTELdirect cooperates with an Internet marketing agency that ensures placing in relevant search engines.

On the website there are 307 German hotels, with last minute availability data. The breakdown by price range is as follows, Euros per room a night in a single room, breakfast included:

Over 80 Euros   127   41%

50 to 80 Euros	123	40%
Under 50	57	19%
Total	307	100%

There are 70 - or 23% of the 307 hotels that are online bookable. This is thanks to cooperation with Worldres.com. The breakdown by price range of the online bookable hotels is as follow

Over 80 Euros	42	33% of high price hotels
50 to 80 Euros	25	20% of medium range hotels
Under 50	3	5% of low price hotels
Total	70	23% of the 307 hotels, in average

It is mainly the high-price hotels which are bookable online, namely one in three. In the medium range it is one in five and in the low price category it is one in 20.

On the web site, as well as on the WAP site and other media, many of the hotels feature information about whether or not the individual hotel has rooms available:

- tonight,

- tomorrow night, and the

- night after tomorrow night.

If we look at 'tonight', checked between four and five in the afternoon, the result is the following, for the 307 hotels:

Available rooms tonight	138 hotels	45%
No available rooms tonight	70 do.	23%
No information about availability	99_ do.	32%
Total	307 hotels	100%

In total, availability information is given for 45+23=68% of the 307 hotels, i.e. for about two thirds. In average 23% of the hotels are explicitly fully booked 'tonight'. There are variations between cities, with respect to how booked out the hotels are. In Nurnberg, some event seemed to be going on, 26 Nov. 2002, then the check was undertaken: Of 17 hotels, as many as 13 or 76% were fully booked between 4 and 5 p.m., on the particular date. In most, i.e. in 11 out of the 13 major cities there were at least some hotels that were fully booked 'tonight'. Only in the two cities with fewest hotels listed, there were no fully booked hotels 'tonight', between 4 and 5 p.m..

The number of hotels listed for each of 13 major cities range from 61 in Berlin to 5 in Dusseldorf and 5 in Rostock. One top of this comes 30 other towns with 78 hotels in total, i.e. 2 or 3 on average for these additional towns. Berlin accounts for 20% of all the 307 German hotels in HOTELdirect, in November 2002. There are also many hotels listed from Hannover (9%); Bremen (8%); Hamburg (7%); Nurnberg and Frankfurt a.m. (6% each); Aachen, Cologne, Munich (4% each); 4 other major cities (7%); 30 additional towns (25%).

Apart from the 307 German hotels in the system there are more than 30000 hotels worldwide/outside of Germany, which can be booked via

HOTELdirect's website. The website is powered by worldres.com, i.e. for all the hotels outside of Germany in the system, the hotel information including availability data and booking function comes from worldres.com.

8.19.2 The WAP site

HOTELdirect launched its WAP site in at ITB in March 2000. The same 307 hotels are displayed on WAP as on the web site. The hotels are listed by city/town, price range and location in relation to railway station or airport. As on the website it is stated whether each of the listed hotels have rooms available tonight, tomorrow night and the night after tomorrow night. It is possible to call any of the hotels by clicking the call function. It is not necessary to provide any personal data before accessing the hotel information on WAP.

HOTELdirect is mainly used by business people and last minute bookers. WAP is not used by overseas visitors coming to Europe. The WAP site was developed in-house, in association with i2dm consulting & development gmbh, also based in Bremen.

In the beginning of 2000 WAP was an entirely new market area, a new platform. There are many users at the WAP site, according to HOTELdirect. One of the reasons for this is that ADAC (the German automobile owners' association) uses hoteldirect as their hotel booking system in its WAP site, i.e. HOTELdirect is one of nine info-services listed at ADAC's WAP site. There is an active hyperlink at the ADAC WAP site, taking those interested in hotel accommodation straight to HOTELdirect's WAP site. HOTELdirect count the number of connections (visitors) to its WAP site.

In a normal hotel reservation system there is commission of 10% or so. However, in the case of HOTELdirect the hotel should pay 2 Euros (only), irrespective of number of nights. Payments from the hotels to HOTELdirect are based on number of calls per month, assuming a certain (high) call-to-book ratio.

HOTELdirect look forward to UMTS, which will enable the swift transmission of pictures to any 3G devise. HOTELdirect work closely with Deutsche Telekom who has development expertise in Bremen.

8.20 Usability case studies conclusion

More than a dozen hotel-related players in the European market had developed WAP sites by early 2001. Some held that WAP on 2G failed because of lack of content. However this series of case studies showed that many hotel players succeeded with the technology on 2G, second-generation mobile networks and the corresponding initial WAP-enabled mobile phones. The case studies were based on personal interviews and all followed the same format: 1) The firm, 2) its web site, 3) its WAP site.

For some of the players the general impression was that they developed the WAP site just to demonstrate that they were capable of utilising the WAP technology and that they were perceived to be at the forefront technologically. Examples include TIScover and Lastminute.com. In other instances the WAP sites were developed because the developers truly believed in the potential of WAP. However, the actual usage of the WAP sites was quite modest in most cases, the ratio of WAP site visits to website visits were no better than 1:20 in favour of fixed-line and in some cases less than 1:1000 - This corresponds well with findings in other travel services such as train travel, where the ratio between WAP and Web usage is typically 1:200 for time-table enquiries. In a couple of instances, the WAP site was developed and run by a dedicated WAP-player, i.e. someone whose main focus is WAP. Each of these players (CityWiz and Cityguides.ch) have a fixed-line website for purely cosmetic reasons and for providing background information about their respective companies.

Chapter 9

9.1 Bringing it all together: WAP usability best practices

Throughout the course of this book we have demonstrated that a careful consideration of interaction and usability issues will result in a more useful small screen user experience.

9.1.1 – Some common issues with WAP

Some of the core issues with WAP result from the following constraints:

- Limited screen size.
- Navigation and input methods.
- Site structure.

9.1.2 – Best practices

Provide users with the information they require quickly

Accessing a WAP site is a totally different experience to accessing a website, users are confined to a very small display and a limited input mechanism and they are likely to be paying for their Internet access by the second. Therefore information needs to be made available quickly without the user having to mine your site for it. Where possible you should minimise the need the user to navigate between multiple cards to get to information and only provide necessary, pertinent information within a card.

Clearly display and distinguish menu items

As outlined earlier, some of the feedback on our prototype was that the menu items were not displayed clearly enough and that some multi-line links were difficult to distinguish as being a single link. It is very important with such a limited interface to ensure that menu items are clearly indicated to save the user frustration at selecting an incorrect link or having to spend unnecessary time trying to decipher what exactly the links do. It is also very important to clearly highlight selected items within your menus, if a user cannot easily tell what link is selected the user experience becomes very frustrating. You can clearly distinguish selected links within a menu by using a different font style and size.

Only provide necessary information

With limited screen real estate and limited processor power, memory and bandwidth the WAP site content must be scaled down to the bare minimum. You must carefully undertake a content audit and identify exactly what your audience requires from the WAP site and remove all of the unnecessary information.

Use dynamic menus, don't repeat them on every card

A standard website features a static menu for navigation, however with your WAP site you will be working within a very confined area and cannot afford to have content pushed off screen by a menu which repeats at the top of every card. Your menus should be context-sensitive and appropriate to where the user is within the deck otherwise the user is going to suffer from losing valuable real estate to unnecessary links.

Ensure user input is simplified

Where possible users should be selecting links within your site to make choices. As mobile phones often have fiddly input mechanisms you should not be asking users to enter large amounts of free text and should certainly not be creating large forms for users to complete on the mobile device. Where possible site registration should be removed or massively downscaled in order to keep the user experience a positive one.

Indicate primary actions clearly

Text and links on WAP sites often look very similar and can tend to blend into each other, any primary navigation links such as 'Submit' or 'Next' should be clearly identified with a different font style, size and location to aid user input and define a clear path to completion.

Chapter 10 – Conclusion

This book has taken you through the process of learning about WAP and WML, looking at how you can go about building a WAP site and more importantly the importance and process of ensuring your WAP site is usable.

The case studies section provides excellent real world information on how companies that are early adopters of WAP technology have found the experience. As you will have noted many of these WAP sites are now closed due to low numbers of visitors, which was mainly attributed to either limitations of the device, the network or prohibited costs of accessing the mobile Internet.

It is important to bear in mind that although WAP is unlikely to be the technology that carries forward the mobile Internet, the mobile Internet **is certainly here to stay**. WAP can almost be viewed as an early test run. What we do learn from WAP is new ways of looking at usability on mobile devices, we are forced to take into account limited screen space and bandwidth, a small amount of CPU power and memory and an awkward text input system.

It is quite likely that in the future, as devices become progressively more powerful and improved 3G network technologies are introduced we will be accessing the 'full web' using these devices or at least a specially adapted version of the full web based on the HTML specification rather than WML. This will actually be a good thing, it is quite apparent that WML does not have sufficient security features in place to support secure transactions, which severely restricts its usage as a tool for purchasing products and services.

The future is looking bright for mobile devices, as we move toward a mobile Internet usability will be an even more important issue than it was with the full flavour web, we will be forced to critically evaluate what information needs to be on a page, to strip unnecessary graphics and images and to use the most efficient methods of presenting navigation and hierarchy. When you combine this improved model of usability with exciting advances such as Location Based Services and GPS, the future is looking very bright indeed and if nothing else WAP will always be remembered as the technology that got us started.

WAP Glossary

A

API: An acronym for Application Programming Interface. The core set of facilities made available to the developer/programmer for writing applications, e.g. system functions and procedures for manipulating information etc.

ASP: ASP is an acronym for Active Server Pages, a Microsoft programming technology that facilitates dynamic content and database integration. ASP is a server-side technology that is used on many thousands of websites. It can be used to create WAP-based applications that need to create content on the fly.

B

Broadband phone: A third generation (3G) mobile phone that has much higher speed access. ETA is 2002.

C

cHTML: cHTML, or compact HTML is a language used to code content in wireless devices, and is used by the popular i-mode system. i-mode is NTT DoCoMo's Internet connection service for mobile phones, and is widely used in Japan, where it has attained some 10 million subscribers. i-mode's cHTML is like HTML 1.0, and is a WML alternative, offering more flexibility and greater features (although it is does not yet use any mature HTML features).

ColdFusion: ColdFusion is an ODBC compliant database integration tool, with its own tag language (CFML) and scripting language (CFMLScript). ColdFusion can be used to write dynamic Web and WAP applications that can serve content on-the-fly. ColdFusion is developed by Allaire.

Cookies: A cookie is a small piece of information stored in memory or on disk. Cookies can be set by client and server-side applications, and are often used as a convenience function, for example to remember users site-login credentials, or to auto-fill a form.

D

DSN: A Data Source Name is a piece of information that links an embedded database query to a specific database on a server. DSNs are stored in files on the users host server. DSNs are not always required; ASP allows "DSN-less" connections for example, which are extremely helpful because the server administrator must set up DSNs. An unresponsive ISP can therefore delay the activation of your database if the DSN is not in place.

DTD: Acronym for Document Type Definition. A DTD definition states which elements can be nested within others, and acts as a rule-set, defining the names and contents of all elements that are allowed within a document, their order and quantity etc.

E

ECMA: ECMA is an international, Europe-based industry association founded in 1961 and dedicated to the standardization of information and communication systems.

ECMAScript

ECMAScript is an attempt to standardize the JavaScript and Jscript (Microsoft) scripting language technologies that are used extensively in both client- and server-side environments. ECMA is a standards setting body, based in Switzerland.

Element: An element specifies the markup and structural information inside a WML deck. Some elements are termed containers, in that they have start and end tags such as the <p> and </p> (paragraph) tags, whilst others exist by themselves e.g. the
 (line break) tag.

G

GSM

An acronym for Global System for Mobile Communication. GSM is the name of the standard around which nearly all mobile networks currently operate. GSM is to be replaced by UMTS sometime in 2002.

GPRS

GPRS is an acronym for General Packet Radio Service, a packet-based wireless communication service that brings data rates from 56Kbps to 114 Kbps, and a continuous connection to the Internet (termed as an "always-on mode") for mobile phone and computer users. The higher data rates offered by GPRS will allow users to participate in video conferences and access multimedia web content. GPRS is based on Global System for Mobile (GSM) communication and will complement existing services such circuit-switched cellular phone connections and the Short Message Service (SMS). GPRS is not a replacement for WAP, since GPRS is a transport-level protocol, whereas WAP is concerned with data, security and mark-up of mobile content taken from the Internet. GPRS is associated with the third generation (3G) mobile phones, many of which are expected to arrive on the market in Q4 2000. GPRS trials are already taking place and the technology is expected to become dominant.

H

HDML

HDML (Handheld Devices Markup Language) - now called the Wireless Markup Language (WML) - is a language that allows portions of HTML to be presented on cellular mobile phones and Personal Digital Assistants (PDAs) via wireless access. Developed by Unwired Planet, HDML amd WML are now open languages.

HTML

Acronym for HyperText Mark-up Language. A tag-based language of elements that perform mark-up (appearance) operations on text within a document.

HTTP

The Hypertext Transfer Protocol (HTTP) is the set of rules for exchanging files (text, graphic images, sound, video, and other multimedia files) on the World Wide Web. Relative to the TCP/IP suite of protocols (which are the basis for the exchange of all data on the Internet), HTTP is an application-based protocol.

HTTP Header

Information sent back by a HTTP web server, including details such as UA and cookies, etc.

I

IIS

An acronym for Internet Information Server, Microsoft's web server product. IIS is a popular industry strength web server, supporting technologies such as ASP, VBScript and JScript.

Infinite loop

See Loop, infinite.

ISP

An acronym for Internet Service Provider, a company that primarily offers Internet access (and other services) to consumers.

J

JavaScript

A scripting language technology used in the main web browsers (Netscape Navigator and Microsoft Internet Explorer) as a client-side technology, and also as a server-side integration tool. Microsoft's alternative is termed JScript, and isECMAScript compliant. WMLScript is similar to JavaScript in terms of syntax, although with a different object model (since the web browser and WAP device environments differ widely).

JScript

JScript is Microsoft's JavaScript implementation, and is fully ECMAScript compliant. JScript can be used as a client and server-side development language, and can be integrated with other Microsoft technologies such as ASP.

L

Loop, infinite

See Infinite loop.

M

M-commerce

A term referring to Mobile Commerce, a hybrid of e-commerce. Mobile commerce is effectively the ability to conduct monetary transactions via a mobile device, such as a WAP enabled cell phone. M-commerce is seen as the Holy Grail of the wireless device market.

MIP

An acronym, first used by Forrester Research, standing for Mobile Internet Provider. MIPS are analogous to ISPs, although they are dedicated to providing wireless services. Some ISPs will merge with MIPs to provide both web and wireless web services.

O

ODBC

Open Database Connectivity (ODBC) is a standard or open application-programming interface (API) for accessing a database. Through using ODBC statements in a program one can access files in a number of different formats, including Access, dBase, Excel, and delimited text. A separate ODBC driver is required for each vendor's database, and a variety of pre-written drivers are available. ODBC is based upon the Open Group standard known as Structured Query Language (or SQL) Call-Level Interface (CLI), which allows applications to use SQL queries that access databases without having to know the proprietary interfaces to those databases. ODBC provides a "mapping" function, taking the SQL request and converting it into a request for data that the individual database system understands.

P

PHP

A hypertext-processing language, with ODBC database integration features. PHP is an open source language, and is used widely on the Internet. It is similar in operation to ASP. More information on PHP (plus download) can be found online at http://www.php.net.

PDA

An acronym for Personal Digital Assistant. A hand-held device such as the Palm Pilot. WAP works with a range of mobile devices and not just cell phones.

_____

S

SDK

An acronym for Software Development Kit. SDKs are available from many mobile operators, such as Nokia and Ericsson. They facilitate the development of applications for wireless devices using technologies such as WML and WMLScript.

Simulator

A simulator is a term used to describe a WAP device implemented in software and as such not a physical device. Simulators are provided as part of many SDKs and allow for local WAP development, saving time spent on air with a real WAP device testing applications. Simulators can also be used with online content, by connecting directly to a given website using HTTP.

SGML

An acronym for Standard Generalised Markup Language, the standard from which HTML was borne.

SMS

An acronym for Short Message Service. A messaging service supported by many mobile phones that allows short text messages, typically in the region of 120 characters, to be sent between mobile devices. SMS can be used to configure WAP phones and to send icons (bitmaps) to phones. SMS is not an interactive protocol, like WAP, which allows interactions with web-based content (using the HTTP protocol).

SQL

An acronym for Structured Query Language, a standard way of accessing and updating information in a database. SQL is database-independent, and uses ODBC to allow it to be used across a variety of different database environments. Many programming environments support SQL, such as ColdFusion and ASP, amongst many others.

T

TCP

TCP (Transmission Control Protocol) is a method (or protocol) used in conjunction with the Internet Protocol (IP) to send data in the form of message units (datagrams, or packets) between computers over the Internet. Whilst IP takes care of handling the actual delivery of the data, TCP takes care of keeping track of the individual units of data that a message is divided into for efficient routing through the Internet. TCP is known as a connection-oriented protocol, which means that a connection is established and maintained until such time as the message or messages to be exchanged by the application programs at each end have been properly exchanged.

U

UA

Acronym for User Agent. A user agent is another name for a WAP device, or web browser, that interprets content coded in formats such as WML, WMLScript, HTML etc. WAP device and microbrowser are other terms for a UA.

UDP

UDP (User Datagram Protocol) is a communications method (protocol) that offers a limited amount of service when messages are exchanged between computers in a network that uses the Internet Protocol (IP). UDP is an alternative to the Transmission Control Protocol (TCP). Like TCP, UDP uses the Internet Protocol (IP) to actually get a data unit (called a datagram, or packet) from one computer to another. Unlike TCP, however, UDP does not provide the service of dividing a message into packets (or datagrams) and reassembling it at the other end, hence the termconnectionless.

UMTS

UMTS, or Universal Mobile Telephone System is the name of a new mobile networking standard that will replace GSMsometime in 2002. UMTS has data speeds many hundreds of times faster than GSM and will bring a true multimedia experience to mobile phones (much faster than the GPRS system that will arrive before UMTS).

URL

An acronym for Uniform Resource Locator. URLs are addresses of web- and WML-based based resources, and can refer to static pages and to applications (scripts). URLs can load both local and remote content, and are made up of a service type, a hostname, and an optional pathname.

V

VBScript

VBScript is a Microsoft scripting (development language) technology that is based on Microsoft's highly successful Visual Basic language. VBScript can be used in the client, and on the server. See also JavaScript, JScript, andECMAScript

---

W

W3C

Acronym for the World Wide Web Consortium, the Web's main standards body. See http://www.w3c.org.

WAE

An acronym for Wireless Application Environment. Specified by the WAP Forum, the Wireless Application Environment defines a general-purpose application environment based fundamentally on web technologies and specifies an environment that allows operators and service providers to build applications that can reach a wide variety of different wireless platforms. WAE is part of the WAP standard.

WAP

WAP is the Wireless Application Protocol. A specification for a set of communication protocols to standardise the way that wireless devices, such as cellular mobile telephones and PDAs can be used for Internet-based access. Four companies conceived WAP, namely: Ericsson, Nokia, Motorola, and Unwired Planet (now Phone.com). WAPs protocol layers are as follows:

Wireless Application Environment (WAE)

Wireless Session Layer (WSL)

Wireless Transport Layer Security (WTLS)

Wireless Transport Layer (WTP)

WAP Device

A WAP device is any device (e.g. mobile phone, PDA, or simulator) that allows access to wireless (WML) content.

WAP Gateway

A WAP gateway is a two-way device, with the WAP device on one side, and the web server on the other. The task of the WAP gateway is to convert content into that suitable for a WAP device. On the web server's side the gateway can provide additional information about the WAP device through items such as HTTP headers.

WAP Server

This term is used in different contexts, by different mobile operators. Essentially, a WAP server is analogous to a web server, e.g. a machine that serves web content according to the HTTP protocol. Most WAP servers are HTTP servers. Some mobile operators have WAP servers that also have a gateway facility as well, notably Nokia's WAP server, allowing the serving of web content, and the serving of WML. WML content is tokenised into an internal format before delivery to a WAP device, a process which is handled inside the WAP server (in the WAP gateway).

WASP

Acronym for Wireless Applications Service Provider, an organisation that provides content and applications for wireless devices, but not necessarily the technical infrastructure (like a MIP does, for example).

WML

WML (Wireless Markup Language), formerly called HDML (Handheld Devices Markup Language), is a tag language that allows the text portions of Web pages to be presented on cellular phones and Personal Digital Assistants (PDAs) via wireless access. WML is used for delivering data to WAP devices, and is HTML-like in its appearance. An alternative to WML is I-Mode's cHTML language.

WML Card

A card is a single block of WML code, which can contain basic text or navigation items. Each card makes up part of the interface for a WML-based application. WML cards must exist within a WML deck, and all WML decks conform to theXML standard.

WML Deck

A collection of WML cards. An entire deck is first loaded whenever the user or WAP device requests a URL.

WMLScript

A scripting language for use with WAP devices. Based on ECMAScript, like JavaScript, but less capable.

WSP

An acronym for Wireless Session Protocol. A Nokia specified technology; WSP provides the upper-level application layer of WAP with a consistent interface for two session services. The first is connection-orientated service like TCP, that operates above a transaction layer protocol, and the second is a connectionless service (like UDP) that operates above a secure or non-secure datagram transport service.

WTAI

An acronym for Wireless Telephony Application Interface. The WTAI specification describes standard telephony-specific extensions to WAE, including WML and WMLScript interfaces to such items as call control features, address book and phonebook services. WTAI was devised by the WAP Forum, and is not universally supported.

X

XML

An acronym for Extensible Markup Language. The W3C's standard for Internet Markup Languages. WML is one such language, and is a subset of SGML. XML describes the structure of content, unlike HTML that describes how pages are "marked-up", i.e. how they appear when viewed with a suitable UA.

Index